Hydraulic Systems Volume 6

Troubleshooting and Failure Analysis

Dr. Medhat Kamel Bahr Khalil, Ph.D., CFPHS, CFPAI.
Director of Professional Education and Research Development,
Applied Technology Center, Milwaukee School of Engineering,
Milwaukee, WI, USA.

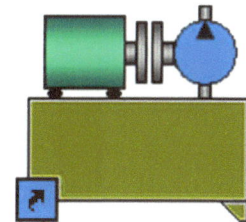

CompuDraulic LLC
www.CompuDraulic.com

CompuDraulic LLC

Hydraulic System Volume 6

Troubleshooting and Failure Analysis

ISBN: 978-0-9977634-6-1

Printed in the United States of America
First Published by June 2022
Revised by ----------------------

Disclaimer

It is always advisable to review the relevant standards and the recommendations from the system manufacturer. However, the content of this book provides guidelines based on the author's experience.

Any portion of information presented in this book might not be suitable for some applications due to various reasons. Since errors can occur in circuits, tables, and text, the author/publisher assumes no liability for the safe and/or satisfactory operation of any system designed based on the information in this book.

The author/publisher does not endorse or recommend any brand name product by including such brand name products in this book. Conversely the author/publisher does not disapprove any brand name product not included in this book. The publisher obtained data from catalogs, literatures, and material from hydraulic components and systems manufacturers based on their permissions. The author/publisher welcomes additional data from other sources for future editions. This disclaimer is applicable for the workbook (if available) for this textbook.

Hydraulic Systems Volume 6
Troubleshooting and Failure Analysis

Chapter 4: Troubleshooting and Failure Analysis of Pumps, 68

PREFACE

Troubleshooting and failure analysis are very important experience to resolve hydraulic systems problems and to find the root causes of such problems. Gaining such experience help to avoid future unexpected shutdowns, hence improve system reliability. This book introduces the approach of logic and analytical troubleshooting fault detection methodologies.

This book is targeting industry professionals who are in charge for operating, maintaining, and troubleshooting hydraulic systems. This book is also a great resource for mechanical engineers and service manuals technical writers.

The book presents more than 40 troubleshooting charts to cover system-level and components-level troubleshooting including hydraulic fluids, pumps, motors, valves, cylinders, accumulators, reservoirs, transmission lines, heat exchanges, filters, and sealing elements. The book also contains proposed inspection sheets for the aforementioned components and investigations for the typical types of failures for each component.

The author is working hard to finish his goal of supporting fluid power professional education by developing the following series of volumes and relevant software:

- Hydraulic Systems Volume 1: Introduction to Hydraulics for Industry Professionals.
- Hydraulic Systems Volume 2: Electro-Hydraulic Components and Systems.
- Hydraulic Systems Volume 3: Hydraulic Fluids and Contamination Control.
- Hydraulic Systems Volume 4: Hydraulic Fluids Conditioning. Under Development
- Hydraulic Systems Volume 5: Safety and Maintenance.
- Hydraulic Systems Volume 6: Troubleshooting and Failure Analysis.
- Hydraulic Systems Volume 7: Modeling and Simulation for Application Engineers.
- Hydraulic Systems Volume 8: Design Strategies of Hydraulic Systems. (Under Development).
- Hydraulic Systems Volume 9: Design Strategies of Electro-Hydraulic Systems. (Under Development).
- Hydraulic Systems Volume 10: Hydraulic Components Modeling and Simulation. (Under Development).

Dr. Medhat Kamel Bahr Khalil

ACKNOWLEDGEMENT

This book was written during the hardship of Covid-19 Virus.

All praises are to Allah who granted me the knowledge, resources and health to finish this work

To the soul of my parents who taught me the values of believe in God

To my wife who offered me all the best she can to make this work completed

To my family: wife, sons, daughters in law, and grandson "Adam"

To my best teachers and supervisors

The author also thanks the following gentlemen for their effective support in developing this book:
- Kamara Sheku, Dean of Applied Researches at Milwaukee School of Engineering.
- Tom Wanke, CFPE, Director of Fluid Power Industrial Consortium and Industry Relations at Milwaukee School of Engineering.

The author also thanks the following companies (listed alphabetically) for granting the permission to use portions of their copyrighted literatures in this book.

- American Technical Publishers
- Assofluid
- Bosch Rexroth
- C.C. Jensen Inc
- Donaldson
- Fluid Power Safety Institute
- Gates Corporation
- Hydraulic and Pneumatic Magazine
- Insane Hydraulics
- Liebherr
- Noria Corporation
- Pall Corporation
- Parker Hannifin
- Spectro Scientific
- Wandfluh
- Womack

Lastly, the author extends his thanks to the following sources of public information used to enrich the contents of the book.

- www.alamo-industrial.com
- www.bonvepumps.com
- shop.finaldriveparts.com
- http://info.texasfinaldrive.com
- metroforensics.blogspot.com
- www.tractorbynet.com
- www.fluiddynamics.com

ABOUT THE BOOK

Book Description:

This book is targeting industry professionals who are in charge of operating, maintaining, and troubleshooting hydraulic systems. This book is also a great resource for mechanical engineers and service manuals technical writers. The book presents more than 40 troubleshooting charts to cover system-level and components-level troubleshooting including hydraulic fluids, pumps, motors, valves, cylinders, accumulators, reservoirs, transmission lines, heat exchanges, filters, and sealing elements. The book also contains proposed inspection sheets for the aforementioned components and investigations for the typical types of failures for each component. This book is colored and has the size of standard A4. The book is associated with a separate colored workbook. The workbook contains printed power point slides, chapter reviews and assignments. This book is the sixth in a series that the author plans to publish to offer complete and comprehensive teaching references for the fluid power industry. The book contains a total of fourteen chapters distributed over 250 pages with very demonstrative figures and tables. The contents of the book are brand non-biased and intends to introduce the latest technologies related to the subject of the book.

Book Objectives:

Chapter 1: Hydraulic Systems Troubleshooting Logical Methodology
This chapter discusses the common methodologies applied for hydraulic system fault detection. This chapter introduces, in a step-by-step, the logic methodology for hydraulic system troubleshooting.

Chapter 2: Basic Troubleshooting Equipment
Servicing staff for hydraulic-driven machines should be aware of the troubleshooting equipment that are used in detecting faults of hydraulic systems. This chapter presents examples of troubleshooting equipment for hydraulic systems.

Chapter 3: Troubleshooting and Failure Analysis of Sealing Elements
This chapter presents guidelines for inspecting and troubleshooting hydraulic sealing elements. The chapter also presents 26 different failure modes, their causes and suggested solutions.

Chapter 4: Troubleshooting and Failure Analysis of Pumps
This chapter discusses hydraulic *pumps* inspection, troubleshooting, and failure analysis. In this chapter, troubleshooting charts for twelve different faults of hydraulic pumps are presented. The chapter also presents examples of defective pumps due to contamination, overheating, cavitation, and fatigue stress for gear, vane, and piston pumps.

Chapter 5: Troubleshooting and Failure Analysis of Motors
This chapter discusses hydraulic *motors* inspection, troubleshooting, and failure analysis. In this chapter, a troubleshooting chart for motor faults is presented. The chapter also presents examples of defective motors due to various reasons such as contamination, clogged case drain, shaft failure, etc.

Chapter 6: Troubleshooting and Failure Analysis of Cylinders
This chapter discusses hydraulic *cylinders* inspection, troubleshooting, and failure analysis. In this chapter, a troubleshooting chart for cylinder faults is presented. The chapter also presents examples of defective cylinder due to various reasons such as contamination, improper mounting, improper load attachment, side loading, overpressure, overheating, fluid incompatibility, saltwater, external leakage, etc.

Chapter 7: Troubleshooting and Failure Analysis of Valves
This chapter discusses hydraulic *valves* inspection, troubleshooting, and failure analysis. In this chapter, a troubleshooting chart for valve faults is presented. The chapter also presents examples of defective hydromechanical and electrohydraulic valves due to various reasons such as particulate and chemical contamination, solenoid burning due to inrush current, etc.

Chapter 8: Troubleshooting and Failure Analysis of Accumulators
This chapter discusses hydraulic *accumulators* inspection, troubleshooting, and failure analysis. In this chapter, a troubleshooting chart for accumulator faults is presented. The chapter also presents examples of defective accumulators caused by various reasons such as vessel explosion due to material defects and pressure shocks etc.

Chapter 9: Troubleshooting and Failure Analysis of Reservoirs
This chapter discusses hydraulic *reservoirs* inspection, troubleshooting, and failure analysis. In this chapter, a troubleshooting chart for reservoirs faults are presented. The chapter also presents examples of defective reservoirs.

Chapter 10: Troubleshooting and Failure Analysis of Transmission Lines
This chapter discusses hydraulic *transmission lines* inspection, troubleshooting, and failure analysis. In this chapter, a troubleshooting chart for transmission line faults is presented. The chapter also presents examples of defective transmission lines.

Chapter 11: Troubleshooting and Failure Analysis of Heat Exchangers
This chapter discusses hydraulic *heat exchangers* inspection, troubleshooting, and failure analysis. In this chapter, a troubleshooting chart for heat exchanger faults is presented. The chapter also presents examples of defective heat exchangers.

Chapter 12: Troubleshooting and Failure Analysis of Filters

This chapter discusses hydraulic *filters* inspection, troubleshooting, and failure analysis. In this chapter, a troubleshooting chart for filter faults is presented. This chapter also presents examples of defective filter.

Chapter 13: Hydraulic Systems Troubleshooting

This chapter introduces troubleshooting charts for failures of generic hydraulic systems. Each troubleshooting chart includes relevant notes and examples for better understanding.

Chapter 14: Examples of Hydraulic Systems Troubleshooting

In this chapter several case studies are presented as examples of applying the logic trouble shooting methodology for hydraulic systems fault detection. In addition, troubleshooting case studies following analytical fault detection methodology are presented. Examples were chosen from both industrial and mobile applications.

Book Statistics:

Chapter #	Pages	Figures	Tables	Words	Editing Time (min)
Chapter 1	7	2	3	1557	8798
Chapter2	13	15	0	1428	8786
Chapter 3	35	49	3	4270	5429
Chapter 4	40	53	14	4828	6606
Chapter 5	8	5	2	1021	5452
Chapter 6	13	16	2	1569	5737
Chapter 7	18	17	5	3132	5759
Chapter 8	6	5	2	1065	5442
Chapter 9	5	4	2	472	5251
Chapter 10	11	16	2	1462	5569
Chapter 11	7	7	2	883	5367
Chapter 12	5	4	2	549	5264
Chapter 13	41	30	18	8091	5582
Chapter 14	46	32	2	11681	6473
Total	255	255	59	42008	85515/60 = 1,425 Hour = 60 Days

ABOUT THE AUTHOR (IFPS Hall of Fam Reciepient)

Medhat Khalil, Ph.D. is Director of Professional Education & Research Development at the Applied Technology Center, Milwaukee School of Engineering, Milwaukee, WI, USA. Medhat has consistently been working on his academic development through the years, starting from bachelor's and master's Degrees in Mechanical Engineering in Cairo Egypt and proceeding with his Ph.D. in Mechanical Engineering and Post-Doctoral Industrial Research Fellowship at Concordia University in Montreal, Quebec, Canada. He has been certified and is a member of many institutions such as: Certified Fluid Power Hydraulic Specialist (CFPHS) by the International Fluid Power Society (IFPS); Certified Fluid Power Accredited Instructor (CFPAI) by the International Fluid Power Society (IFPS); Member of Center for Compact and Efficient Fluid Power Engineering Research Center (CCEFP); Listed Fluid Power Consultant by the National Fluid Power Association (NFPA); and Listed Professional Instructor by the American Society of Mechanical Engineers (ASME). Medhat has balanced academic and industrial experience. Medhat has vast working experience in Fluid Power teaching courses for industry professionals. Being quite aware of the technological developments in the field of fluid power, Medhat had worked for several world-wide recognized industrial organizations such as Rexroth in Egypt and CAE in Canada. Medhat had designed several hydraulic systems and developed several analytical and educational software. Medhat also has considerable experience in modeling and simulation of dynamic systems using Matlab-Simulink. Medhat has been selected among the inductees for Pioneers in fluid Power by NFPA (2012) and Hall of Fam in fluid Power by IFPS (2021).

Chapter 1
Hydraulic Systems
Troubleshooting Logic Methodology

Objectives

This chapter discusses the common methodologies applied for hydraulic system fault detection. This chapter introduces, in a step-by-step, the logic methodology for hydraulic system troubleshooting.

Brief Contents

1.1- Fault Detection Methodology

1.2- Logic Fault Detection Procedure

1.3- General Component Check

1.4- Noisy Unit

1.5- Excessively Hot Unit

Chapter 1: Hydraulic Systems Troubleshooting Logic Methodology

1.1- Fault Detection Methodology

Troubleshooting is a term applied to a methodical approach of finding out what went wrong in a system and resolve the cause of failure.

Cost of Machine's Down time: Down time on modern production machinery costs a lot. An hour saved in locating a problem can save hundreds, sometimes thousands, of dollars in lost production.

Halfway to System Troubleshooting: As with all *troubleshooting* techniques, knowledge of components and their functions in a system are vitally important. It can be stated that knowing the construction and operating principle of hydraulic components plus knowledge of the machine history are halfway towards finding the problem.

Fault Detection Methodologies: As shown in Fig 1.1, there are two methodologies to determine the cause of a failure in a hydraulic driven machine.

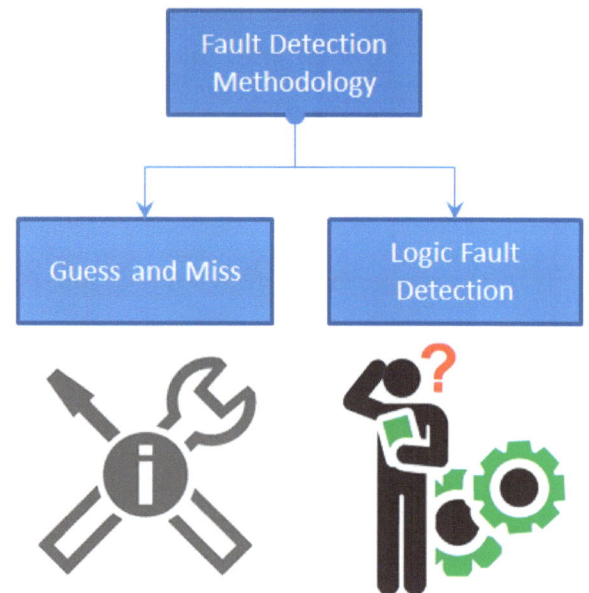

Fig. 1.1- Fault Detection Methodology

Guess and Miss Approach: Units are changed randomly until the failed component is located.
  ❖ **Advantage:** Easiest troubleshooting technique & does not require experienced personnel.
  ❖ **Disadvantage:** It may take long time and large number of perfectly serviceable units are changed before the failed component is found.

Logic (Analytical) Fault Detection Approach: Fault is detected based on series of *logic* steps to find the cause of failure or malfunction of a system or its components.
  ❖ **Advantage:** Cost effective because only failed units will be repaired or replaced.
  ❖ **Disadvantage:** It requires experienced personnel.

1.2- Logic Fault Detection Procedure

A hydraulic system troubleshooter acts like a doctor treating a patient. So, he/she must follow a step-by-step *logic fault detection* procedure.

❖ **Step 1: Review Safety Instructions:**
Before starting the troubleshooting procedure, make sure to review the safety instruction provided by the machine manufacturer, place in which the machine is used, and local municipality. Additionally, review the following best practices in Volume 5 (Safety and Maintenance) of this series of textbooks:

- BP-Safety-03: Safety of Hydraulic System Work Environment.
- BP-Safety-04: Safety of Hydraulic System Workspace.
- BP-Safety-05: Safe Startup of Hydraulic Systems.
- BP-Safety-06: Safe Operation of Hydraulic Systems.
- BP-Safety-07: Safe Servicing of Hydraulic Systems.

Fig. 1.2- Review Safety Instructions before Start

❖ **Step 2: Review Machine History:**
Before analyzing the hydraulic part of the machine, be sure the trouble is not mechanical or electrical. For better fault detection,

- The following information must be reviewed and understood:
  - Machine log.
  - Hydraulic circuit diagram.
  - Control circuit diagram.
  - Latest oil analysis report.

- The following questions must be answered:
  - What is the machine application?
  - Was the development of the fault gradual or sudden?
  - Has the fault occurred after oil changes?
  - Has the fault occurred after change of component change/adjustment?
  - Has the fault occurred previously, how frequently?
  - How does the fault affect the machine operation?
  - How does the fault affect the other connecting processes?

❖ Step 3: Identify Main System Fault:

Based on understanding the machine history, identify the main system fault. The following list summarizes the main root causes of many system-level faults. Each one of these main faults is numbered based on the associated best practices investigation flow chart.

- **T-System-01:** Fluid Aeration.
- **T-System-02:** Pump Cavitation.
- **T-System-03:** Excessive System Noise & Vibration.
- **T-System-04:** Excessive System Heat.
- **T-System-05:** Lack of Load Carrying Capacity.
- **T-System-06:** Faulty System Sequence.
- **T-System-07:** External Leakage.
- **T-System-08:** Troubleshooting Open Hydraulic Circuit.
- **T-System-09:** Troubleshooting Closed Hydraulic Circuit (Hydrostatic Transmission).
- **T-System-10:** Actuator Slow Performance.
- **T-System-11:** Actuator Fast Performance.
- **T-System-12:** Actuator Erratic Performance.
- **T-System-13:** Actuator Moves in Wrong Direction.
- **T-System-14:** Actuator Stops to Move.
- **T-System-15:** Actuator Load Drifts.
- **T-System-16:** Actuator Leaks.

❖ Step 4: Apply the System-Level Troubleshooting Chart:

Among the system-level troubleshooting charts, apply the ones that are assigned to the identified faults. Action items listed in each chart are ordered based on the common causes and the easiness of doing the action. Likely by applying this step the fault will be identified.

❖ Step 5: List Suspicious Components:

If the fault isn't identified so far, in order to narrow the search, list the suspicious components that may contribute in developing the fault. The criteria in listing the suspicious components is to start with the components that are easy to test and easy to access.

❖ Step 6: Perform Preliminary Check on the Suspicious Components:

Inspect each suspicious component using the relevant *inspection sheet*, then apply the preliminary check on each of the suspicious components using the following investigation charts:

- **T-Unit-01:** General Check
- **T-Unit-02:** Noisy Unit.
- **T-Unit-03:** Excessively Hot Unit.

❖ **Step 7: Apply Detailed Check on the Suspicious Components:**
Likely by then, the fault is identified. If not, detailed investigation is required. Review the *Inspection Sheet* and apply the component-level troubleshooting chart for all components listed in the suspicious components list:

- **T-Seal-01:** Seal Troubleshooting.
- **T-Pump-01:** No Flow out of the Pump.
- **T-Pump-02:** Low Flow out of the Pump.
- **T-Pump-03:** Erratic Flow out of the Pump.
- **T-Pump-04:** Excessive Flow out of the Pump.
- **T-Pump-05:** No Pressure at the Pump Outlet.
- **T-Pump-06:** Low Pressure at the Pump Outlet.
- **T-Pump-07:** Erratic Pressure at the Pump Outlet.
- **T-Pump-08:** Excessive Pressure at the Pump Outlet.
- **T-Pump-09:** Leaking Pump.
- **T-Pump-10:** Excessive Pump Wear.
- **T-Pump-11:** Air Leaks into Pump.
- **T-Pump-12:** Excessive Pump Noise and Vibration.
- **T-Valve-01:** DCV Troubleshooting.
- **T-Valve-02:** FCV Troubleshooting.
- **T-Valve-03:** PCV Troubleshooting.
- **T-Valve-04:** EH Valve Troubleshooting.
- **T-Valve-05:** General Valve Troubleshooting.
- **T-Motor-01:** Motor Troubleshooting.
- **T-Cylinder-01:** Cylinder Troubleshooting.
- **T-Accumulator-01:** Accumulator Troubleshooting.
- **T-Reservoir-01:** Reservoir Troubleshooting.
- **T-Transmission Line-01:** Transmission Line Troubleshooting.
- **T-Heat Exchanger-01:** Heat Exchanger Troubleshooting.
- **T-Filter-01:** Filter Troubleshooting.

❖ **Step 8: Fault Evaluation Decision of Repair or Replacement:** A decision must be made to repair or replace a faulty component based on evaluating the fault.

❖ **Step 9: Startup and Testing:**
After removing the cause of fault by repair or replacement, a startup and testing step is required. Before starting up, review safety precautions required for starting up a hydraulic-driven machine from Volume 5 of this series of textbooks.

❖ **Step 10: Future Considerations and Documentation:**
In this last step, take proper actions and list recommendations to assure that this fault won't occur again. It is important to follow up on the system performance, in a period after resolving the problem, to make sure problem doesn't return.

1.3- General Component Check

Any component listed among the suspicious faulty components must pass a *preliminary check* routine. Table 1.1 shows the relevant troubleshooting chart.

T-Unit-01-General Check	
model number or ordering code are incorrect?	▪ Replace the incorrect component based on the correct model number or ordering code.
Is the unit installed correctly?	▪ Follow the guidelines for proper installation of the component.
Does the unit receive a control signal?	▪ Check for proper control signal.
Is the unit adjustable?	▪ Check if the unit is properly adjusted.

Table 1.1- Troubleshooting Chart (T-Unit-01-General Check)

1.4- Noisy Unit

If *noise* is traced to a specific unit, Table 1.2 shows the relevant troubleshooting chart.

T-Unit-02-Noisy Unit	
Is the unit rotating?	▪ Check maximum speed. ▪ Check worn or sticking part
Pressure fluctuations in the return line?	▪ Check for restrictions in return line.
Is the unit coupled to a prime mover or coupled to a rotating mass?	▪ Check the coupling and alignment conditions.
Is the unit adequately supported to sub-plate or pipe works?	▪ Tighten the unit per the recommendation.
Mechanical noise?	Check: ▪ Loose, worn, or misaligned coupling. ▪ Loose set screw. ▪ Badly worn internal parts. ▪ Bearing failure.

Table 1.2- Troubleshooting Chart (T-Unit-02-Noisey Unit)

1.5- Excessively Hot Unit

If *heat* is traced to a specific unit, Table 1.3 shows the relevant troubleshooting chart.

T-Unit-03-Excessively Hot Unit	
Is the unit experiencing internal leakage?	▪ Test the internal leakage and compare it with the allowable rate. ▪ Resolve the issue if found.
Is the unit undersized?	▪ Size the valve properly based on the flow rate.
The unit experiencing high internal friction or seizure?	▪ Check the unit and resolve the issue.
Is the unit rotating or cycling faster than the normal rate?	▪ Adjust the unit performance to the rated value.
Does the case drain restricted and is not cooled?	▪ Resolve restriction and consider cooling case drain.
Is viscosity too low causing lack of lubrication?	▪ Oil may be too thin either from wrong choice of oil or from thinning out at high temperature. Consequently, lack of lubrication causes system overheating and chain action continues.
Is ambient temperature too high?	▪ Consider proper ventilation around the unit.
Is fluid heavily contaminated by abrasives?	▪ Maintain the recommended cleanliness level.
Is the unit covered by dirt?	▪ Keep outside surfaces clean.

Table 1.3- Troubleshooting Chart (T-Unit-02-Excessively Hot Unit)

Chapter 2
Basic Troubleshooting Equipment

Objectives

Servicing staff for hydraulic-driven machines should be aware of the troubleshooting equipment that are used in detecting faults of hydraulic systems. This chapter presents examples of troubleshooting equipment for hydraulic systems.

Brief Contents

2.1- Snap-Check Pressure Gauge Test Kit
2.2- Hydrostatic Transmission Pressure Gauge Test Kit
2.3- Pressure/Leak Test Kit
2.4- Universal Flow Meter Test Kits
2.5- Portable Digital Hydraulic Multimeter
2.6- Adaptor Kit
2.7- Test Points and Pressure Measurement Hoses
2.8- Fluid Leakage Test Kit
2.9- Surface Temperature Thermometers
2.10- Vibration Indicators
2.11- Tachometers
2.12- Multimeters
2.13- Proportional Valve Tester
2.14- Servo Valve Tester

Chapter 2: Basic Troubleshooting Equipment

2.1- Snap-Check Pressure Gauge Test Kit

Figure 2.1 shows an example of a *snap-check* pressure gauge test kit used to test pressure at various points in the hydraulic system.

The HC-TKC1-SC Test Kit includes:
- 0 to 30″ Hg (0 to -1.0 bar) <u>vacuum gauge</u> with 'Snap-Check' <u>diagnostic test coupler</u>.
- 0 to 3000 PSI (0 to 207 bar) <u>pressure gauge</u> with 'Snap-Check' <u>diagnostic test coupler</u>.
- 0 to 6000 PSI (0 to 414 bar) <u>pressure gauge</u> with 'Snap-Check' <u>diagnostic test coupler</u>.
- Two (2) 60″ (1524 mm) <u>microbore hose assemblies</u>. Each hose assembly has one (1) male 'Snap-Check' test nipple and one (1) 'Snap-Check' test coupler on either end.
- One (1) 1/4″ NPT male 'Snap-Check' diagnostic test nipple.
- One (1) 7/16″ ORB male 'Snap-Check' diagnostic test nipple.
- One (1) 9/16″ ORB male 'Snap-Check' diagnostic test nipple.
- One (1) G1/8 male 'Snap-Check' diagnostic test nipple.
- One (1) G1/4 male 'Snap-Check' diagnostic test nipple.
- One (1) M10x1.25 male 'Snap-Check' pressure test connector.
- One (1) M12x1.5 male 'Snap-Check' pressure test connector.
- One (1) M14x1.5 male 'Snap-Check' pressure test connector.

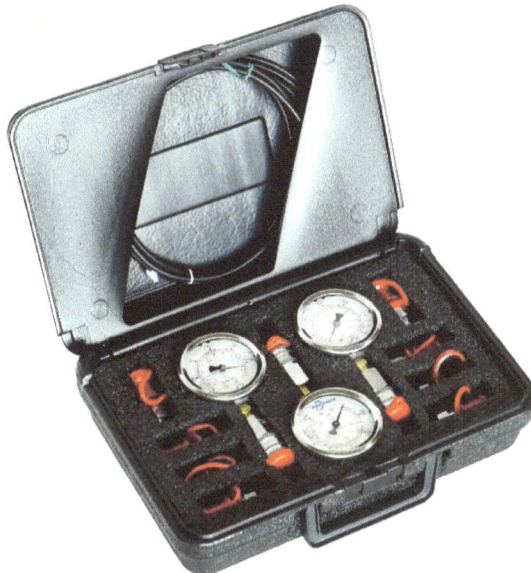

Fig. 2.1- HC-TKC1-SC Test Kit (Courtesy of Hydracheck)

2.2- Hydrostatic Transmission Pressure Gauge Test Kit

Figure 2.2 shows a *hydrostatic transmission* pressure gauge test kit. This special kit is compact and incudes all the pressure gauges and adapters needed to setup, adjust and test most hydrostatic transmission systems.

Hydrostatic Transmission Pressure Gauge Test Kit includes:
- 0 to 100 PSI (0 to 6.9 bar) <u>pressure gauge</u> with direct gauge adapter.
- 0 to 600 PSI (0 to 41.4 bar) <u>pressure gauge</u> with direct gauge adapter.
- 0 to 6000 PSI (0 to 414 bar) <u>pressure gauge</u> with direct gauge adapter.
- Three (3) 60″ (1524 mm) <u>microbore hose assemblies</u>.
- Three (3) microbore <u>hose unions</u> (connect hoses for additional length).
- One (1) 1/4″ NPT pressure test connector.
- One (1) 5/16″ SAE pressure test connector.
- Two (2) 7/16″ SAE pressure test connectors.
- One (1) 9/16″ SAE pressure test connector.

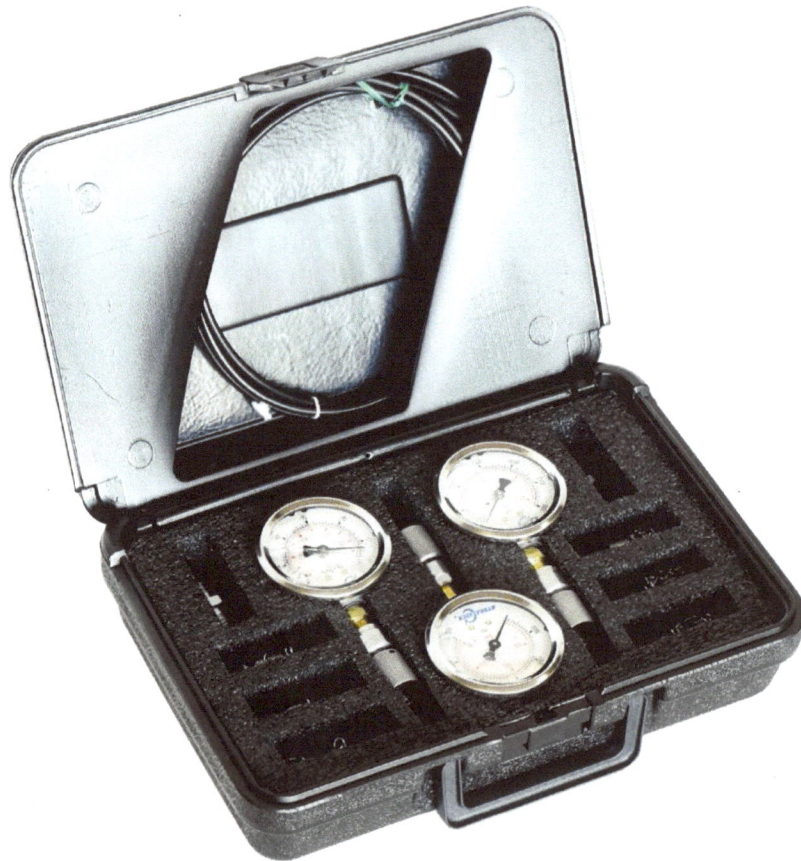

Fig. 2.2- Hydrostatic Transmission Pressure Gauge Test Kit (Courtesy of Hydracheck)

2.3- Pressure/Leak Test Kit

Figure 2.3 shows a *pressure/leak* gauge test kit. This kit allows safe and accurate field-testing of hydraulic components, particularly pressure control valves, check valves, directional control valves, and circuit modules, without ever having to exhaust pressurized oil to atmosphere.

The HC-TK2000 test kit includes:
- 0 to 3000 PSI (207 bar) pressure gauge with direct gauge adaptor.
- 24″ (607mm) hose assembly with STAUFF® Test 15 swivel nut.
- Pressure/leak test pump 3000 PSI (207 bar) maximum operating pressure.
- Infrared, non-contact, mini-thermometer, with laser sighting.
- NPT adaptors sizes 1/4″ to 1-1/2″.
- SAE adaptors sizes #4 to #24.
- JIC adaptors sizes 1/4″ to 1- 1/2″.
- Code 61 adaptors sizes 1/2″ to 1-1/2″.
- Code 62 adaptors sizes 1/2″ to 1-1/2″.
- Optional Caterpillar®-style adaptors sizes 1/2″ to 2″ are available.

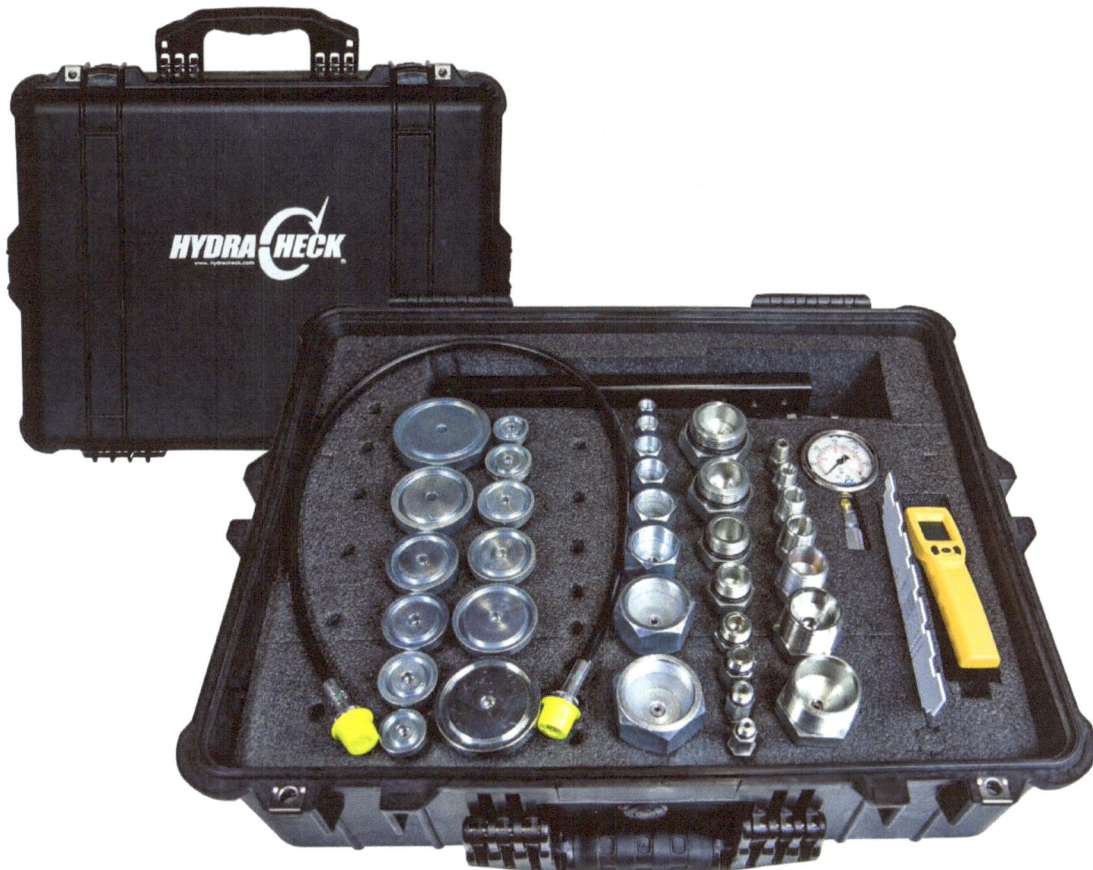

Fig. 2.3- Pressure/Leak Test Kit (Courtesy of Hydracheck)

2.4- Universal Flow Meter Test Kits

To meet the needs of maintenance professionals who service a wide variety of hydraulic machinery, Fig 2.4 shows a *Universal Flowmeter* test kit. This kit covers a wide variety of flows from 1.0 GPM to 100 GPM (3.8 Lpm to 568 Lpm).

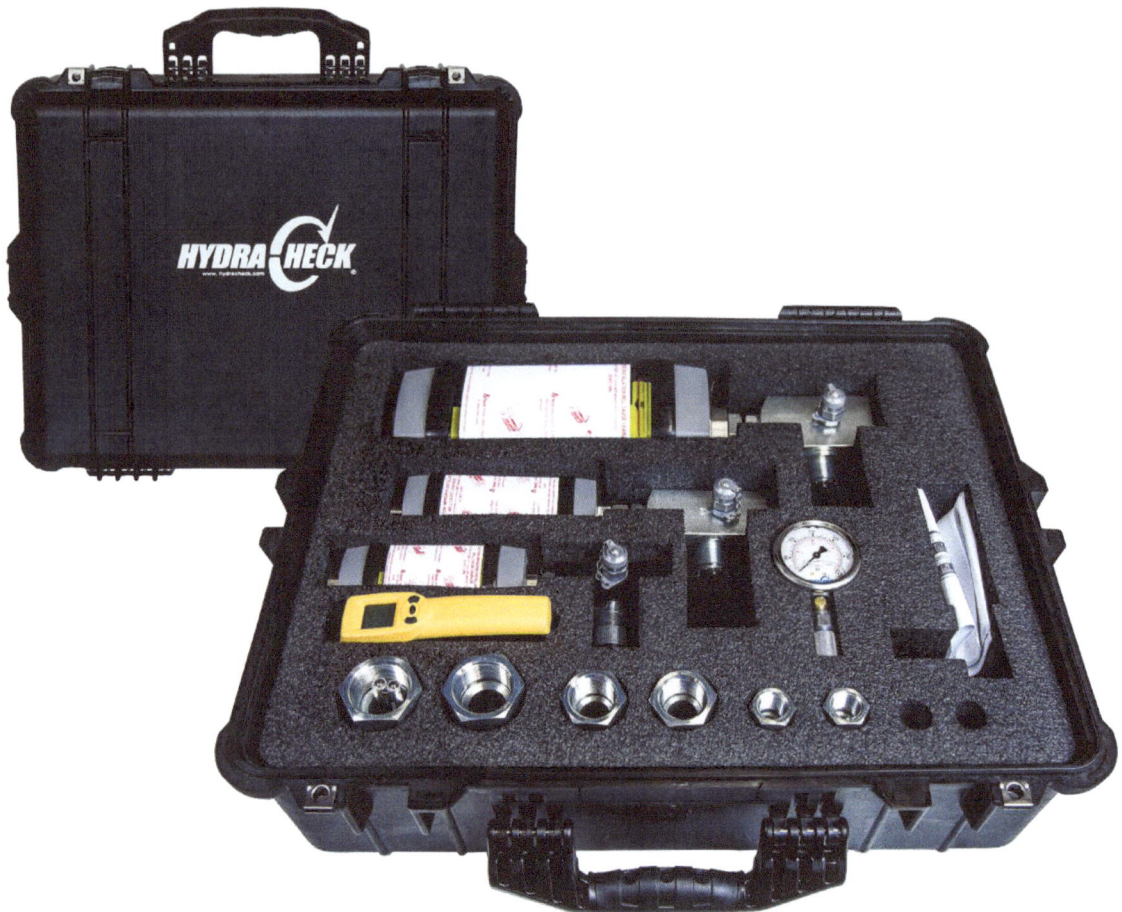

Fig. 2.4- Universal Flowmeter Test Kit (Courtesy of Hydracheck)

2.5- Portable Digital Hydraulic Multimeter

Figure 2.5 shows a *Digital Hydraulic Multimeter* is an all-in-one unit designed to test the performance of hydraulic pumps, motors, valves and hydrostatic transmissions with the added benefit of Bluetooth functionality. This unit can measure hydraulic flow, pressure, peak pressure, temperature, power, and volumetric efficiency and includes onboard memory to record a simple 12 point test, more than sufficient to characterize a hydraulic pump's performance at varying pressures which can be regulated using the built-in loading valve. This new generation of testers, aimed at hydraulic field service personnel everywhere, enables them to easily measure the true performance of hydraulic systems, pinpoint faults, and quickly email a test report in just a few presses of iOS smartphone, using Smart Bluetooth™ functionality. This truly portable solution means the service technician no longer have to return to the office to send his report as all of this can be achieved while working in field.

Specifications:
- Pressures up to 7000 PSI (482 bar).
- Flows up to 210 GPM (950 lit/min).
- Flow Accuracy: ± 1%.
- Pressure Accuracy: ± 0.5% of full scale, Peak 1%.
- Temperature Accuracy: ± 2°F (± 1°C)
- Volumetric efficiency: ± 1%
- 1 ms response time.
- Choose between – Bar, PSI, MPa, & Ksc.

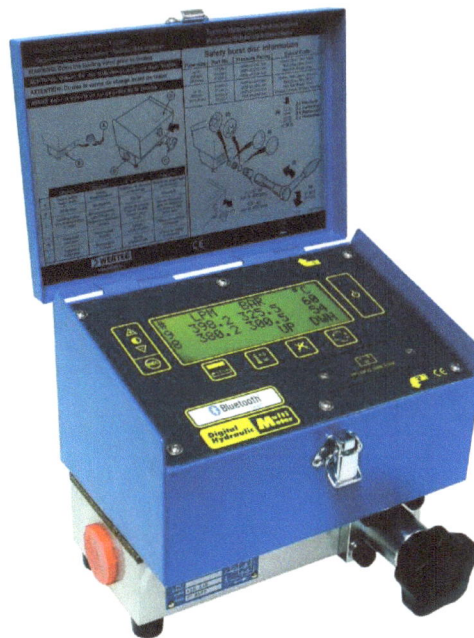

Fig. 2.5- Bidirectional Digital Hydraulic Multimeter with Bluetooth (Courtesy of Hydracheck)

2.6- Adaptor Kit

The *Transition Adapter Kit,* shown in Fig. 2.6, contains all the adapters needed to check pressure on most of hydraulic systems. It can connect to the diagnostic nipples founded on machines manufactured by Caterpillar, John Deere, Komatsu, Volvo, JCB, Genie, etc.

Fig. 2.6- Adaptor Kit (Courtesy of Hydracheck)

2.7- Test Points and Pressure Measurement Hoses

Figure 2.7 shows a set of test points and pressure measurement hoses frequently required in trooubleshooting process. They are available in various sizes and configurations.

Fig. 2.7- Test Points and Measurement Hoses

2.8- Fluid Leakage Test Kit

Figure 2.8 shows a kit for detecting fluid leakage using *Florescent Dyes*. The kit comes in a variety of special formulations to match each specific application. They are available in a range of colors to differentiate between different leaking fluid types. They work with any host fluid without damaging its properties or system components. So, no need to drain the reservoir after detecting the leakage.

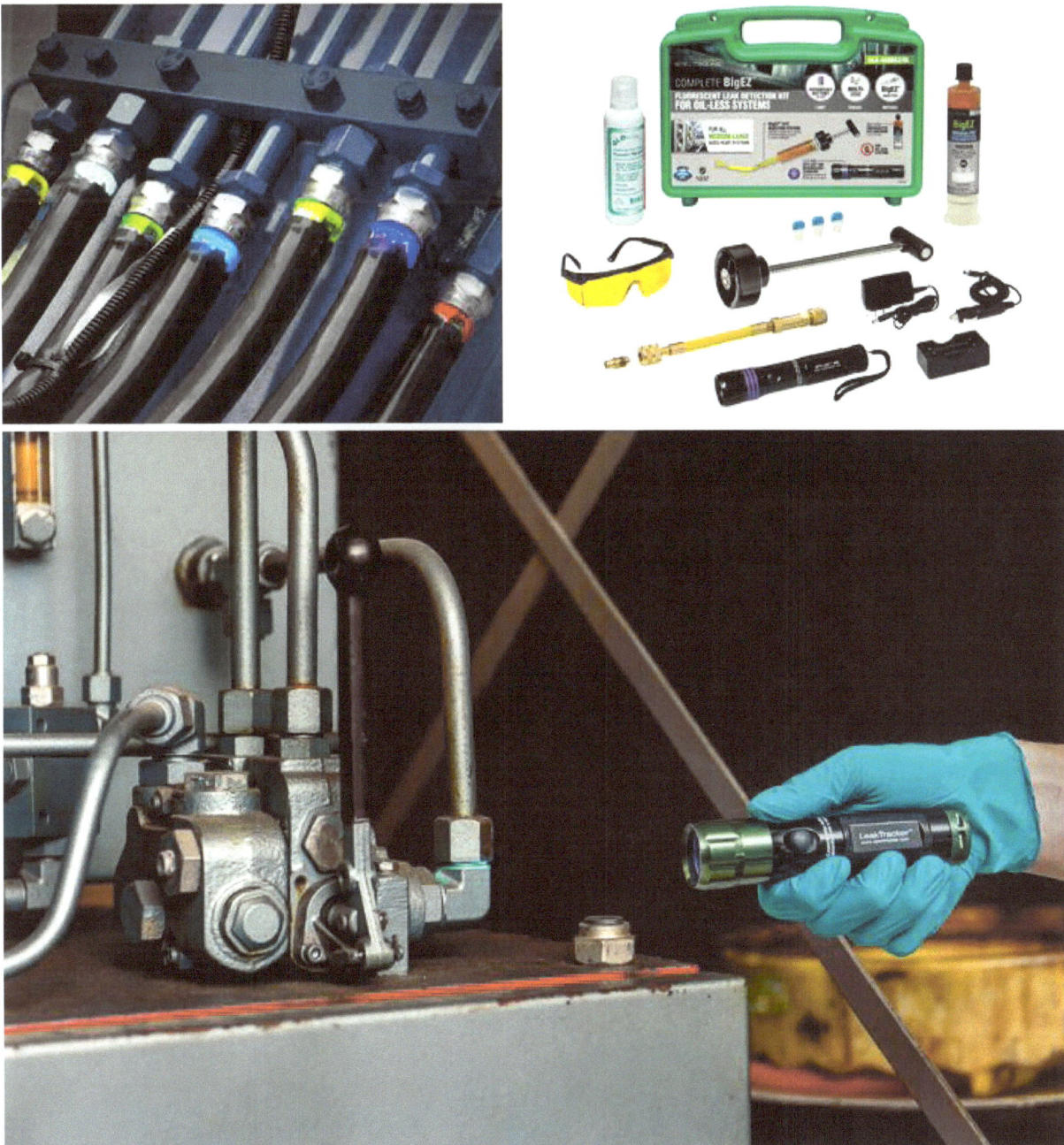

Fig. 2.8- Fluid Leakage Test Kit (Courtesy of Spectroline)

2.9- Surface Temperature Thermometers

Touching hot surfaces is not a good practice because it can result in burning skin. Figure 2.9 shows typical *Surface thermometers*. They are used to measure surface temperature from a distance to detect internal leakage in a hydraulic component or track faults that results in generating heat.

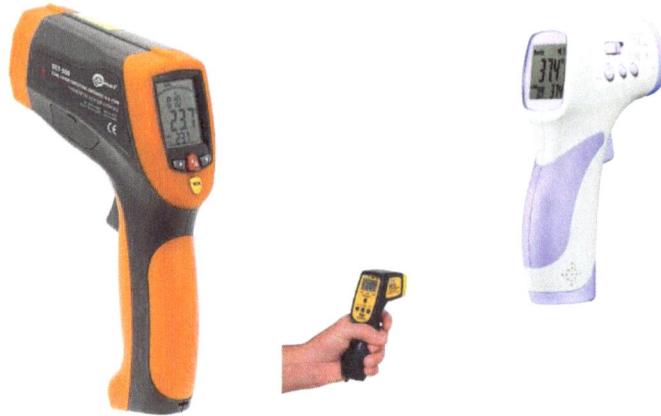

Fig. 2.9- Surface Temperature Thermometers

2.10- Vibration Indicators

To detect mechanical faults of a hydraulic component, *vibration* must be measured. Figure 2.10 shows typical vibration indicators. They are available with a permanently installed vibration sensor or Hand Arm Vibration Indicator (HVAI).

Fig. 2.10- Vibration Indicators

2.11- Tachometers

Checking the rpm of rotating equipment, such as pumps or motors, can be done by permanently installed rpm sensors or *tachometers*. As shown in Fig. 2.11, they are available in different types like mechanical or optical, and contact or contactless.

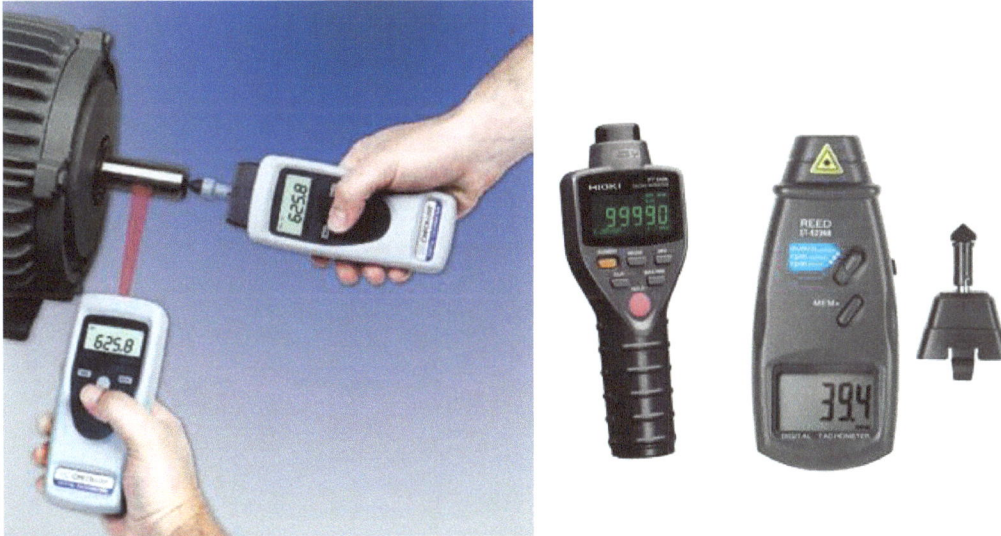

Fig. 2.11- Various Tachometers for Measuring RPM

2.12- Multimeters

Checking electro-hydraulic systems requires measuring Amperes, Volts, Resistances, etc. Figure 2.12 shows typical *multimeter*.

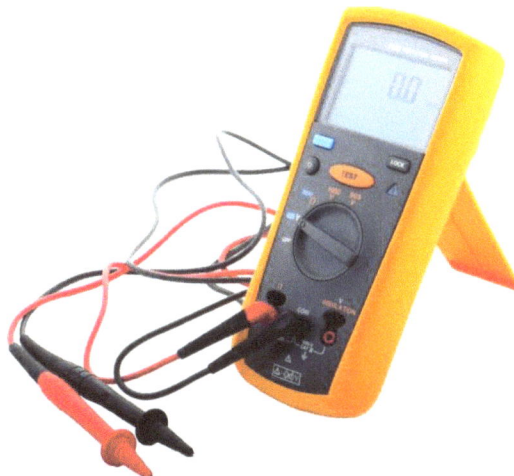

Fig. 2.12- Multimeter

2.13- Proportional Valve Tester

Figure 2.13 shows an example of *Proportional Valve Tester*. Service case contains test device, 24 Volt power supply unit, connection cable and adapter cable. The test device is suitable for the control and functional testing of proportional valves with integrated electronics (OBE) and an operating voltage of ±15 V or +24V. This unit makes commissioning and troubleshooting of hydraulic systems with proportional valves easier.

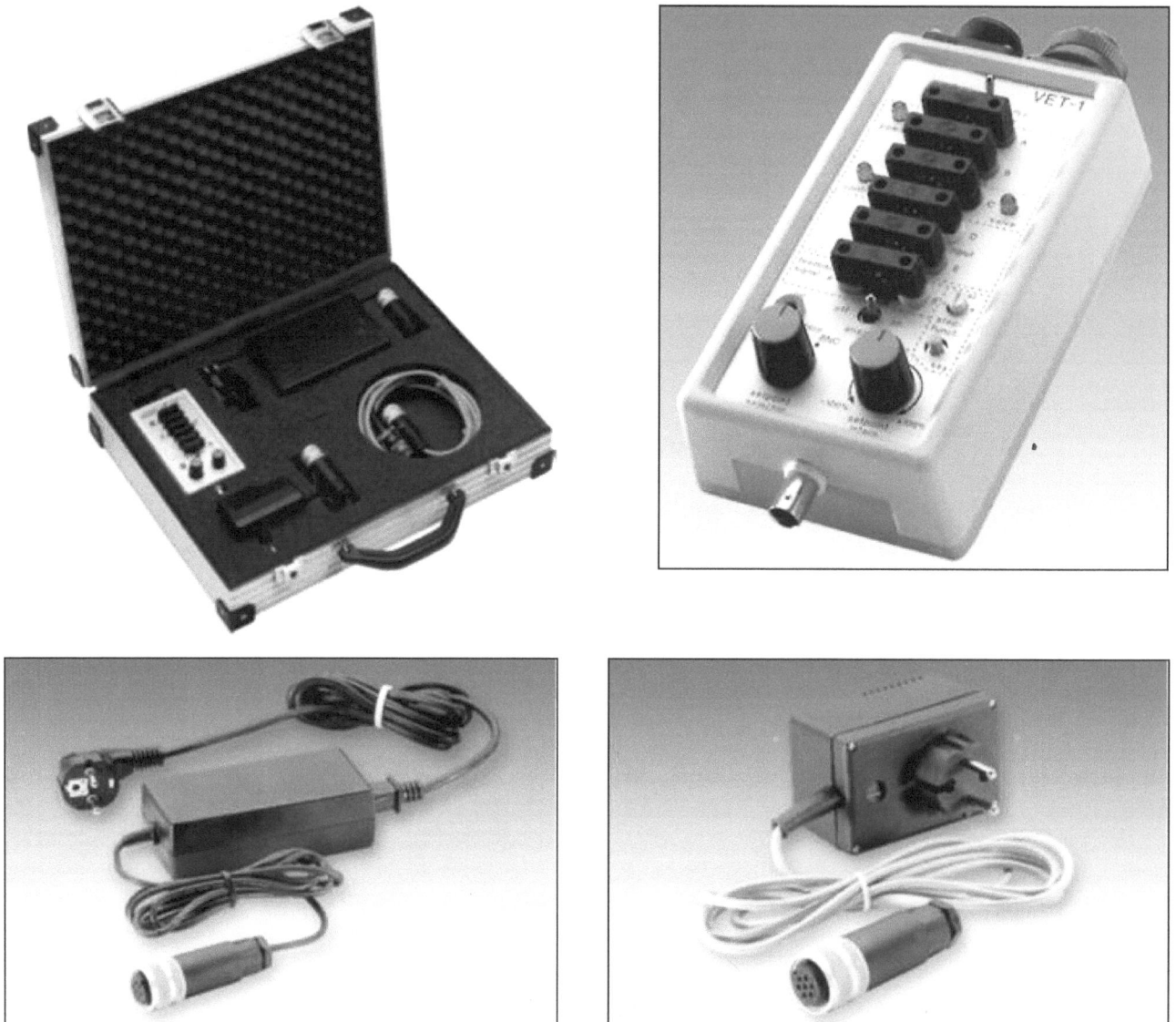

Fig. 2.13- Proportional Valve Tester Type VT-VETSY-1 (Courtesy of Bosch Rexroth)

Figure 2.14 shows another example of *proportional valve tester* that is upgraded with the following features:

- Operation via touchscreen for digital or analog display.
- Automatic valve fault detection.
- Setpoint generator.
- Ramp function.
- Potentiometer operation.
- Two outputs for switching valves.

Caution: These two valve testers should only be used by experienced persons who are familiar with the device, the valve, and the hydraulic system. The use of the device on running systems is always at the user's own risk! The tester manufacturer assumes no liability will be accepted for damage caused by incorrect operation!

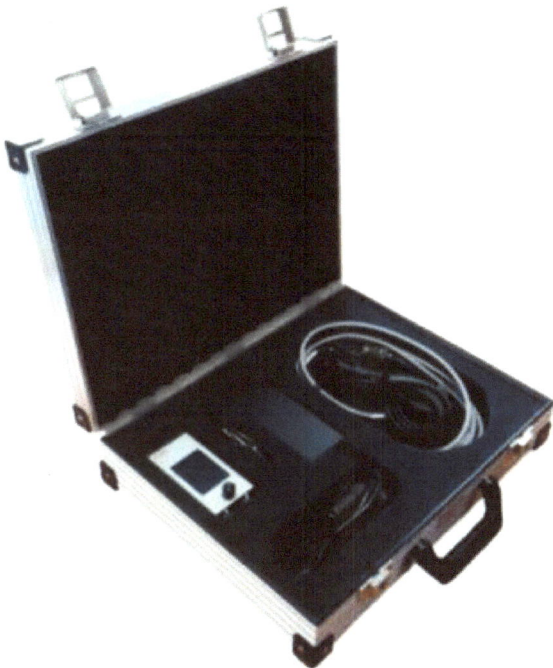

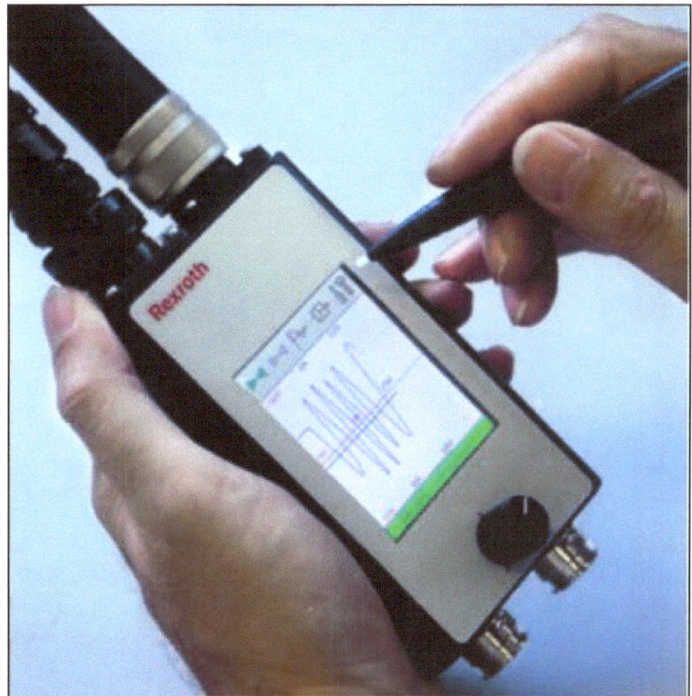

Fig. 2.14- Proportional Valve Tester Type VT-HDT-1-2X (Courtesy of Bosch Rexroth)

2.14- Servo Valve Tester

Figure 2.15 shows an example of *Servo Valve Tester*. The test unit is suitable for testing and commissioning of all proportional and servo proportional valves with onboard electronics. For easy on-site service all necessary cables are securely located inside of the rugged case. The test unit provides all command signals and measuring ports for time saving diagnosis of the valves.

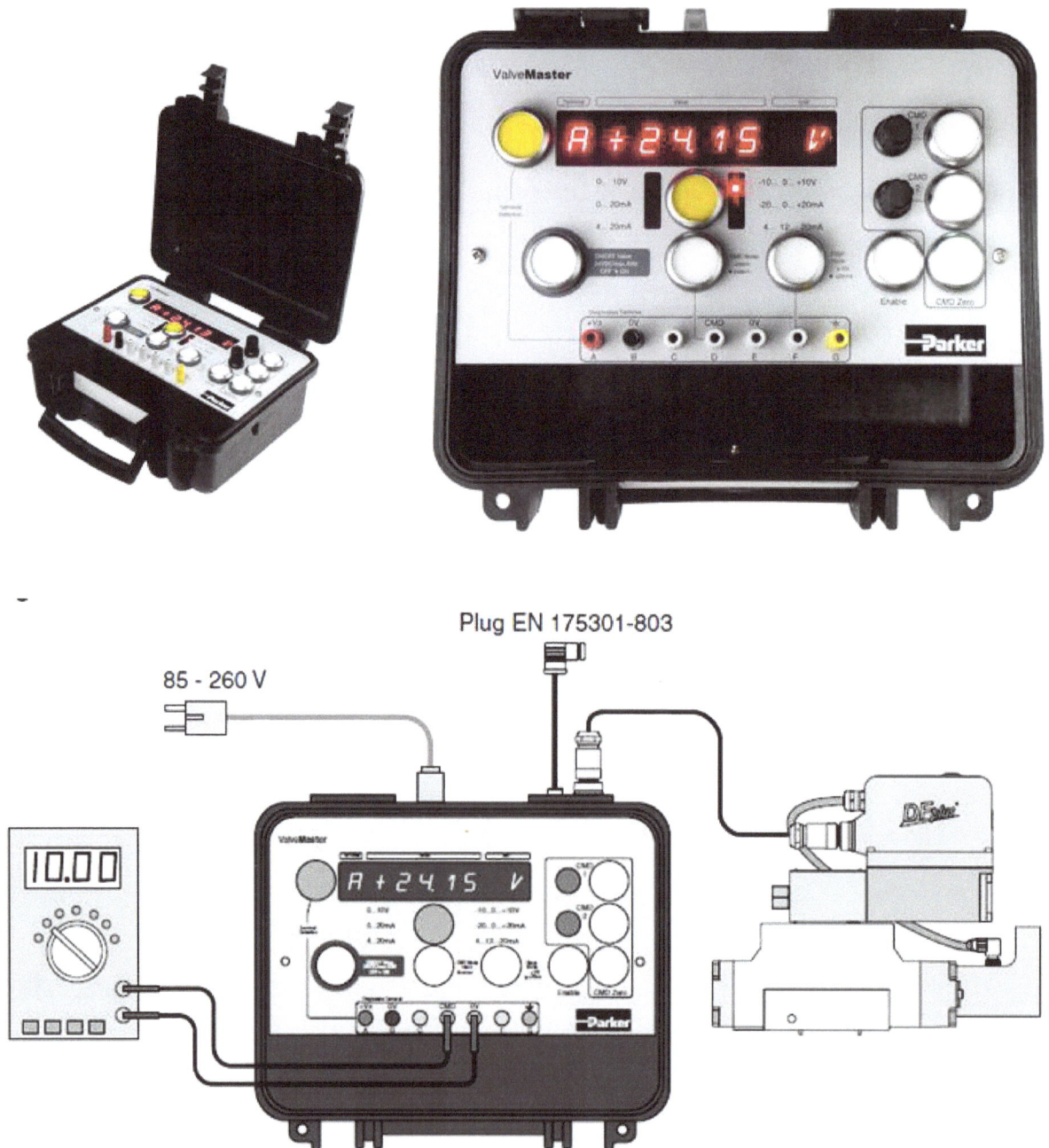

Fig. 2.15- Servo Valve Tester Series EX-M05 (Courtesy of Parker)

Chapter 3

Troubleshooting and Failure Analysis of Sealing Elements

Objectives

This chapter presents guidelines for inspecting and troubleshooting hydraulic sealing elements. The chapter also presents 26 different failure modes, their causes and suggested solutions.

Brief Contents

3.1- Hydraulic Seals Inspection

3.2- Hydraulic Seals Troubleshooting

3.3- Hydraulic Seals Failure Analysis

Chapter 3: Troubleshooting and Failure Analysis of Sealing Elements

3.1- Hydraulic Seals Inspection

Hydraulic *sealing elements* could be as simple as an O-Ring or a complex design of a piston seal package. They could be static or dynamic seals. They serve translational or rotational hydraulic components. Volume 4 of this series of textbooks provides guidelines for sealing elements design, maintenance and safety. Table 3.1 shows a typical inspection sheet for a hydraulic seal.

Hydraulic Seals Inspection Sheet	
Manufacturer	
Model #	
Serial #	
Location	
Type of Seal	☐ Piston Seal ☐ Rod Seal ☐ Rotating Shaft Seal ☐ Other Seal
Seal Failure	☐ Material of a hydraulic seal is harshly scorched. ☐ Seal is severely compressed and deformed in short time. ☐ Seal is extruded. ☐ Seal has abrasion market. ☐ Seal is leaking. ☐ Seal has short cuts. ☐ Seal is crushed and stressed beyond its limits and fails. ☐ Seal is squeezed and failed. ☐ Seal is hardened, became brittle, and has cracks. ☐ Seal is hardened, glazed, and has cracks. ☐ Seal has axial cuts particles embedded in the seal material. ☐ Seal lost its flexibility and cracks are formed. ☐ Seal softened, swell, or shrink. ☐ Seal material break-down, loss of physical properties, cracking, and crumbling. ☐ Seal has signs of bubble decompression. ☐ Seal has signs of Dieseling. ☐ Seal has increased gland clearance on one side only or uneven friction. ☐ Seal has excessive wear. ☐ Seal has torsional or spiral failure. ☐ V-Seal shows long cracks or splits.

Table 3.1 – Hydraulic Seal Inspection Sheet

3.2- Hydraulic Seals Troubleshooting

Table 3.2 shows troubleshooting chart for hydraulic *sealing elements*.

T-Seal-01-Seal Troubleshooting	
Material of a hydraulic seal is harshly scorched.	▪ Improper molding process.
Seal is severely compressed and deformed in short time.	▪ Poor seal material properties
Seal is extruded.	▪ Excessive Sealing Gap. ▪ Overpressure. ▪ Seal material is too soft. ▪ Improper seal size.
Seal has abrasion markes.	▪ Improper surface finish.
Seal is leaking.	▪ Improper seal design. ▪ Seal deterioration. ▪ Side loads.
Seal has short cuts.	▪ Seal pass over sharp edges during installation.
Seal is crushed and stressed beyond its limits and fails.	▪ Seal is over pressurized.
Seal is squeezed and failed.	▪ Improper seal design (pressure trapping).
Seal is hardened, became brittle, and has cracks. Material splits through the seal body.	▪ Seal is overheated. ▪ Normal aging effect.
Seal is hardened, glazed, and has cracks.	▪ Seal is used at high speed.
Seal has axial cuts particles embedded in the seal material.	▪ High abrasive contamination.
Seal lost its flexibility and cracks are formed.	▪ Fluid incompatibility.
Seal softened, swell, or shrink.	▪ Chemical attach.
Seal material break-down, loss of physical properties, cracking, and crumbling.	▪ Exposure to water or emulsions.

36 | Hydraulic Systems Volume 6: Troubleshooting and Failure Analysis
Chapter 3- Troubleshooting and Failure Analysis of Sealing Elements

Seal has signs of bubble decompression.	▪ Fluid aeration. ▪ Sudden decompression of air bubbles.
Seal has signs of Dieseling.	▪ Cavitation. ▪ Sudden decompression of air bubbles.
Seal has increased gland clearance on one side only or uneven friction.	▪ Side loading
Seal has excessive wear.	▪ Rough surface finish of gland. ▪ High working temperature. ▪ Poor fluid lubricity. ▪ Fluid incompatibility ▪ Mechanical vibration. ▪ Abrasive contamination. ▪ Sharp peaks and hard sealing surfaces.
Seal has torsional or spiral failure.	▪ Rough surface finish of gland. ▪ High working temperature. ▪ Poor fluid lubricity. ▪ High stroke speed. ▪ Long stroke. ▪ Side loads. ▪ Squeezed or soft seal. ▪ Too much space for movement in the groove. ▪ Type of metal surface. ▪ ID/W ratio of O-ring. ▪ Contamination or gummy deposits on metal surface. ▪ Eccentricity of sealing ring. ▪ Stretch of sealing rings. ▪ No use of Back-up Rings.
V-Seal shows long cracks or splits.	▪ Seal fatigue due to cold startup and/or exposure to cyclic pressure with high frequency

Table 3.2- Troubleshooting Chart (T-Seals-01-Seals Troubleshooting)

3.3- Hydraulic Seals Failure Analysis

Challenges of Seals Failure:

- **Possible Remedies:** Seal damage is irreversible and there are no remedies for failed seals other than replacement.

- **Consequences of Seal Failure:** The consequences of a seal failure may vary from a simple internal or external leakage to a serious damage. A critical seal would be one that if failure occurs would create hazard for personnel and/or the public. Application examples where critical seals are used are aircraft, amusement park rides, and elevating devices.

- **Cost of the seal vs. the Consequences:** The cost of hydraulic seals is pennies compared to the time and resources put into disassembling components. So, this low cost of seal may result in improper operation of a major component.

- **Failure Analysis:** Without laboratory analysis, it may not be easy by just visual inspection to determine the condition of seal deterioration over the time.

As shown in Fig. 3.1, failure modes of hydraulic sealing elements can be categorized due to the following:

    A. Manufacturing Defects.
    B. Seal and Gland (Groove) Design Issues.
    C. Assembly Procedures.
    D. Operational Conditions.
    E. Normal Aging.
    F. Storage Conditions.

To make it easy to jump to a specific failure mode, the following subtitles are numbered based on the number of the failure mode on the chart.

**Hydraulic Seals
Failure Analysis**

A- Manufacturing Defects

1-Improper Molding

2-Poor Material

B- Design Issues

3-Extrusion

4-Gland Sharp Corner

5-Rough Interface Surfaces

6-Blow-By Effect

C- Assembly Procdures

7-Passing Over Sharp Edges

D- Operational Conditions

8-Fluid Overpressure

9-Pressure Trapping

10-Fluid Overheating

11-Overspeeding

12-Fluid Contamination

13-Fluid Incompatibility

14-Fluid Chemical Attack

15-Hydrolysis

16-Explosive Decompression

17-Dieseling

18-Side Loading

19-Vibration

20-Spiral Failure

21-Seal Wear

22-Fatigue

E- Normal Aging

23-Hardening

24-Splits

F- Storage Conditions

25-Swelling

26-Ozone Cracking

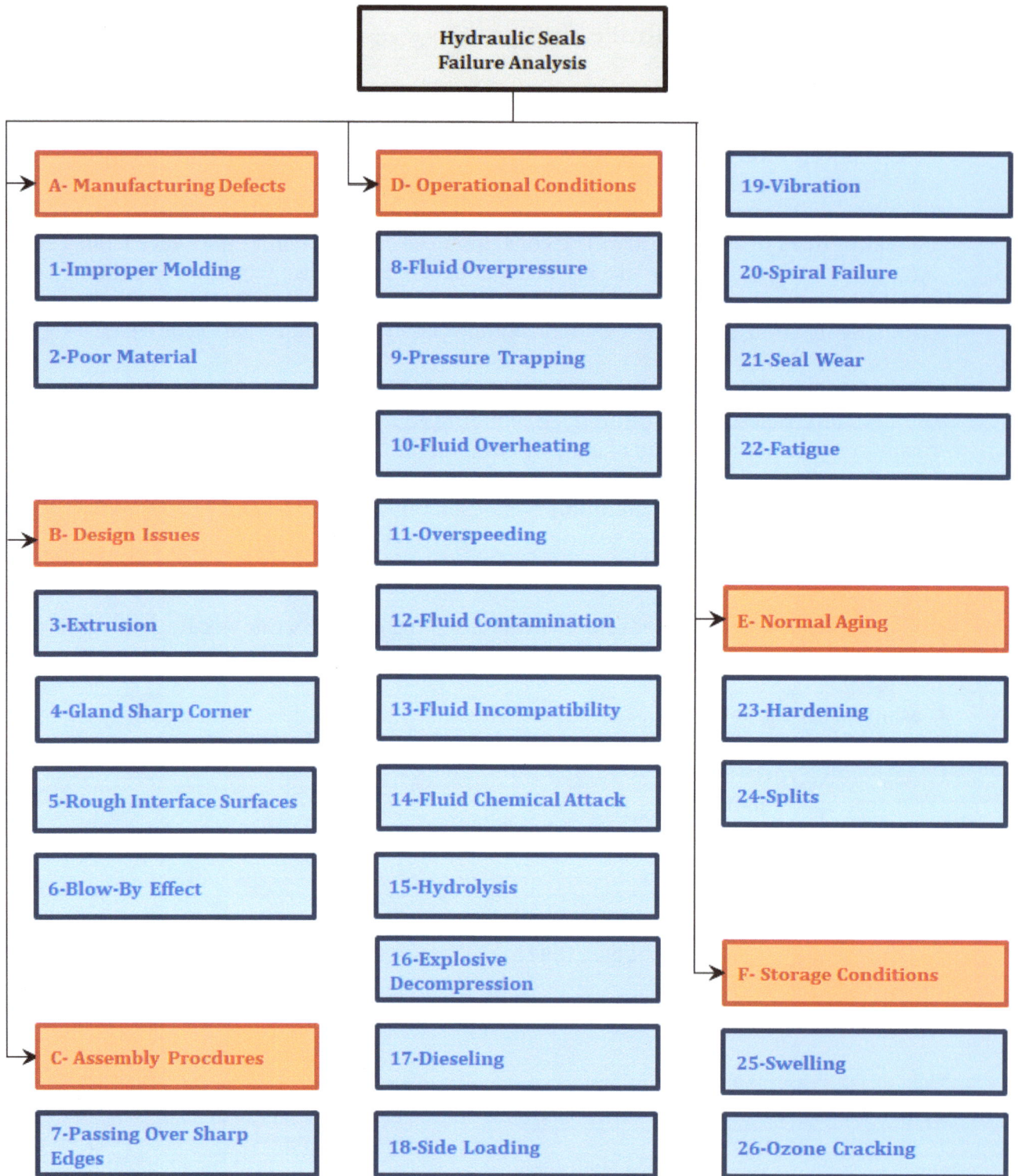

Fig. 3.1 - Hydraulic Seal Failure Analysis Diagram

3.3.1- Manufacturing Issues - Improper Molding

Failure Source: Improper *Molding* process due to defective dies, improper injection flow, pressure or temperature.

Failure Mode: As shown in Fig. 3.2, material of a hydraulic seal is harshly scorched.

Suggested Solution: Review molding process and conditions.

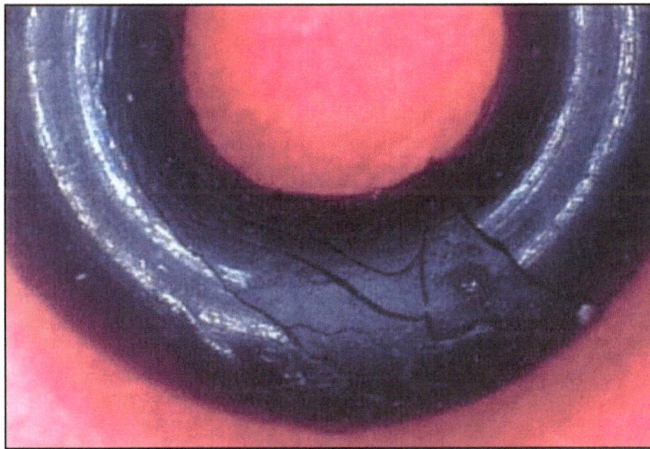

Fig. 3.2 - O-Ring Failure due to Improper Molding (www.o-ring-lab.com)

3.3.2- Manufacturing Defects - Insufficient Material Properties

Failure Source: Severe compression set due to poor material properties.

Failure Mode: As shown in Fig. 3.3, a hydraulic seal is permanently deformed in short time.

Suggested Solution: Select better quality hydraulic seal.

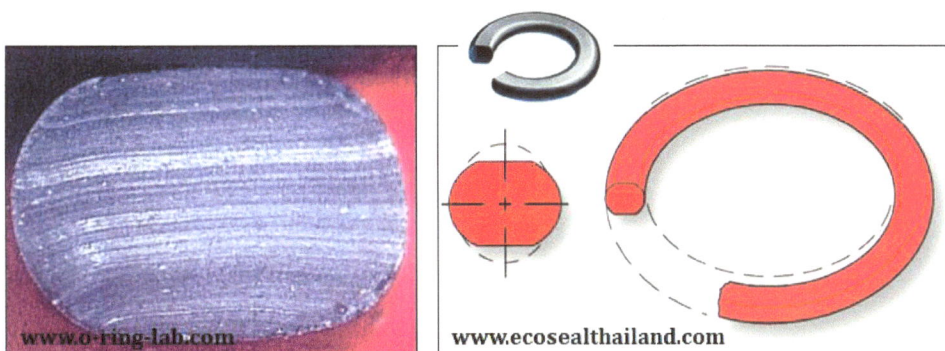

Fig. 3.3 - O-Ring Failure due to Insufficient Material Properties

3.3.3- Design Defects - Extrusion

Failure Source: *Extrusion* and nibbling are caused by one or more of the following conditions:
- **Excessive Sealing Gap:** Caused by excessive clearances due to improper installation.
- **Nonuniform Sealing Gap:** Caused by nonuniform clearance due eccentricity.
- **Less Hardness:** Seal material is too soft.
- **Degradation:** swelling, softening, shrinking, cracking, etc.
- **Improper Size:** Too large seal installed causing excessive filling of groove.
- Other Reasons: Overpressure, overheating, side loading, soft material, and chemical incompatibility.

Extrusion Limits are defined based on the seal hardness "Shore A", diametral clearance, and the working pressure. Figure 3.4 shows the extrusion Limits. As shown in the figure, at the same working pressure (red arrows), the sealing gap (diametral clearance) and extrusion limits are directly proportional to the seal hardness. On the other hand, for the same seal hardness (blue arrows), the sealing gap and extrusion limit are inversely proportional to the working pressure.

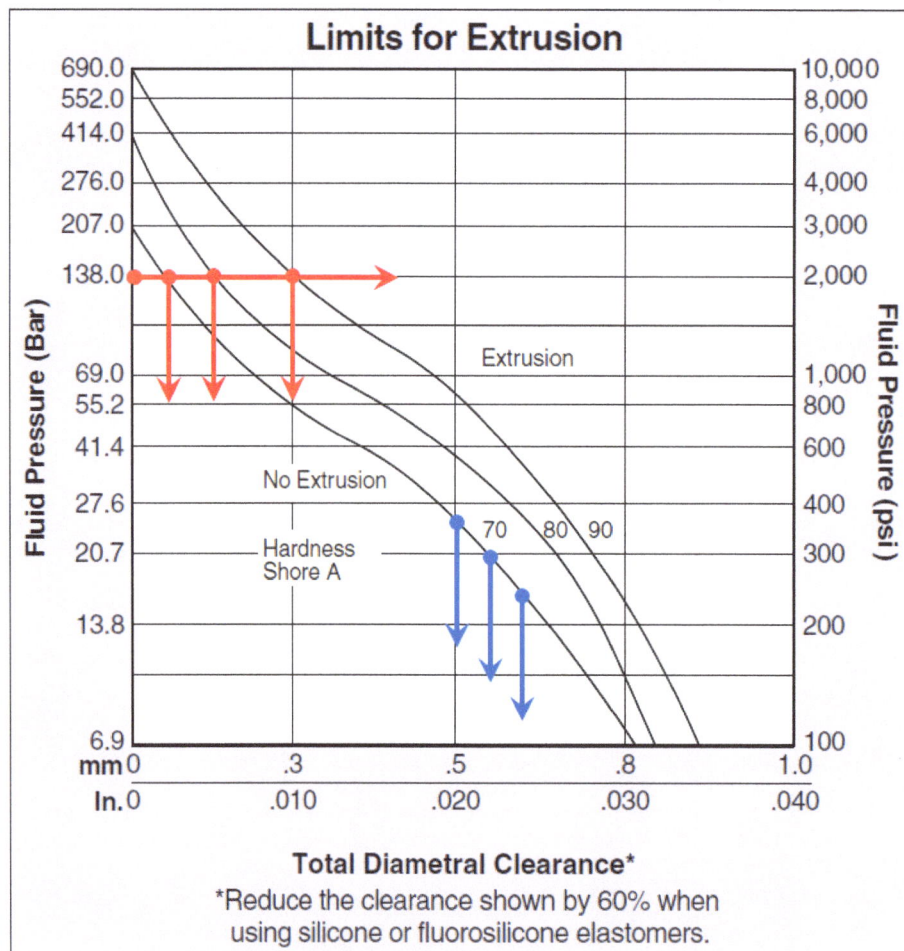

Fig. 3.4 - Limits of Extrusion (Courtesy of Parker)

Failure Mode:

If Extrusion Limits are exceeded, as shown in Fig. 3.5, the O-Ring will extrude into the sealing gap.

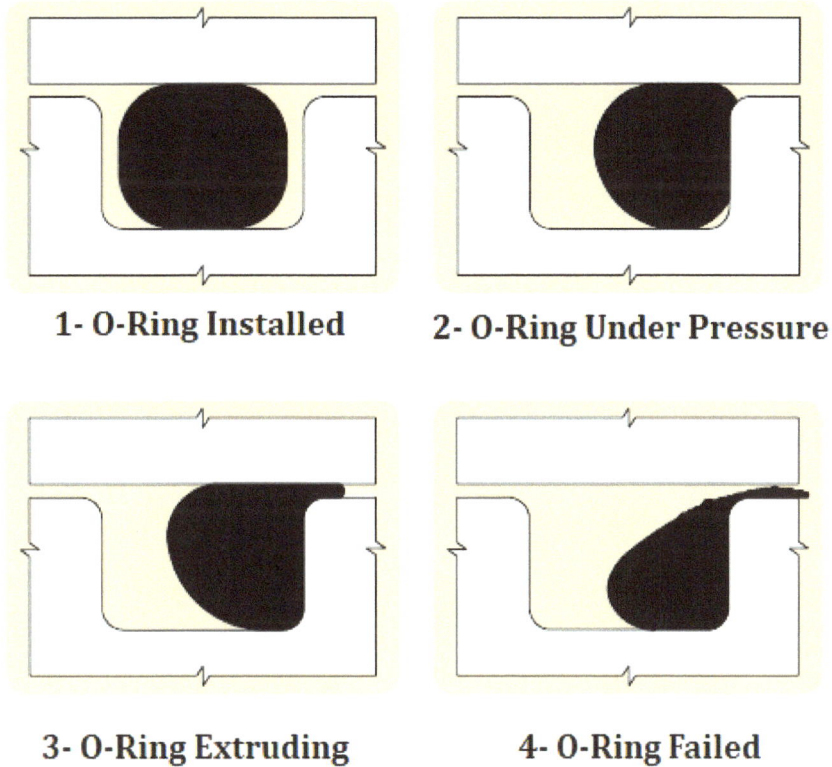

1- O-Ring Installed   2- O-Ring Under Pressure

3- O-Ring Extruding   4- O-Ring Failed

Fig. 3.5 - O-Ring Extrusion due to Excessive Seal Gap (Courtesy of Parker)

Figures 3.6 through 3.8 show hydraulic seals creeping into the sealing gap. As a result, the seal deforms and/or breaks off.

Fig. 3.6 - Seal Extrusion, Example 1 (Courtesy of Parker)

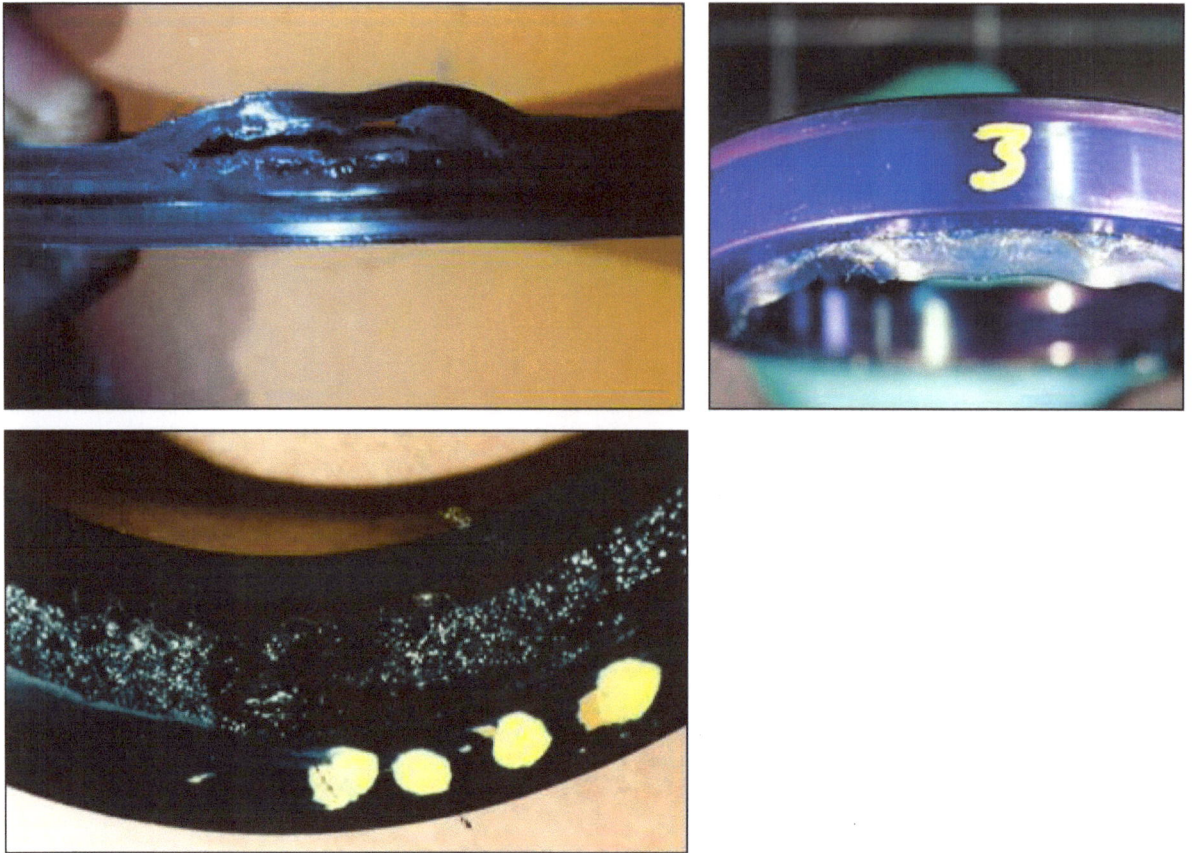

Fig. 3.7 - Seal Extrusion, Example 2

PTFE piston seal showing signs of severe extrusion

Fig. 3.8 - Seal Extrusion, Example 3 (Courtesy of System Seals Inc.)

Standard Test Method - Extrusion-Resistance Test (ASTM C1183 / C1183M): this method is used to determine the ability of a hydraulic seal to resist extrusion under certain pressure.

Suggested Solution: Seal gap extrusion can be avoided by considering the following solutions:

Solution 1-Proper Design of Sealing Gap:

A sealing gap must be properly designed to avoid seal extrusion.

Solution 2-Use of Back-up Rings:
As shown in Fig. 3.9, backup rings are placed on the back side of O-Rings or dynamic seals to support the seal against extrusion.

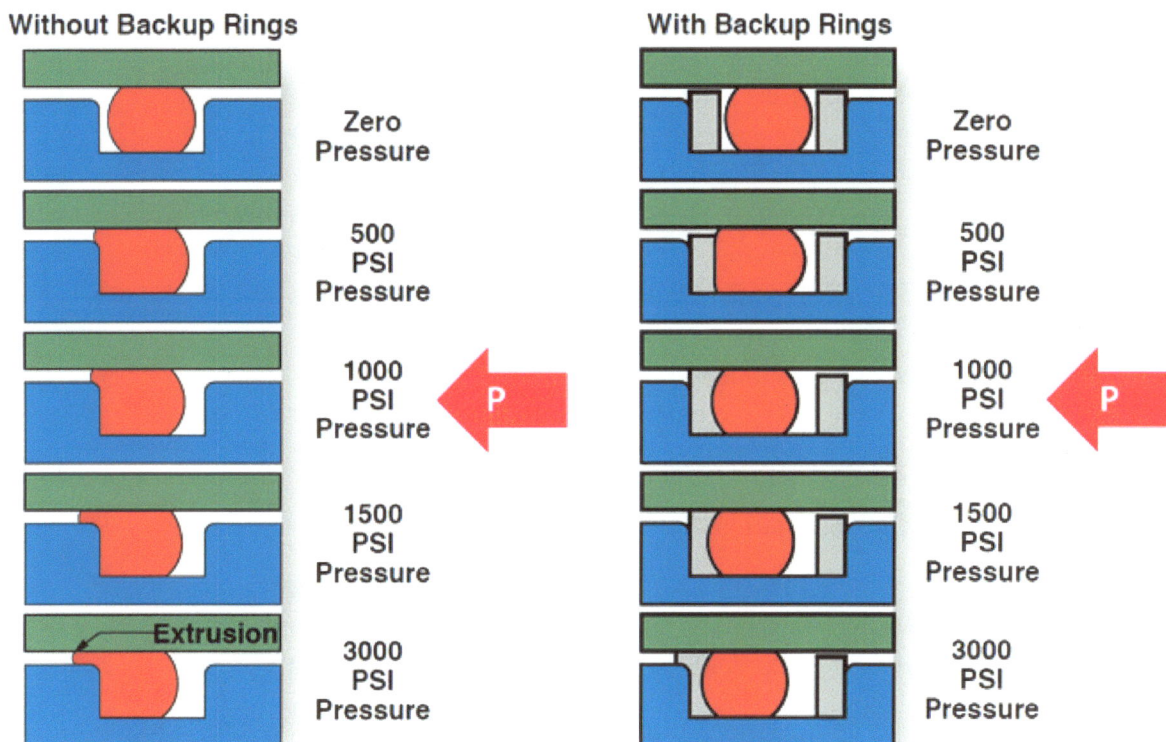

Fig. 3.9 - Seal Extrusion Avoidance by use of Back-up Rings (www.ecosealthailand.com)

Solution 3-Use of Anti-Extrusion Wedge-Rings:
As shown in Fig. 3.10, a *Wedge-Ring* is placed on the back side of an O-Ring. The Wedge-Ring adjusts the gap dynamically as the pressure changes.

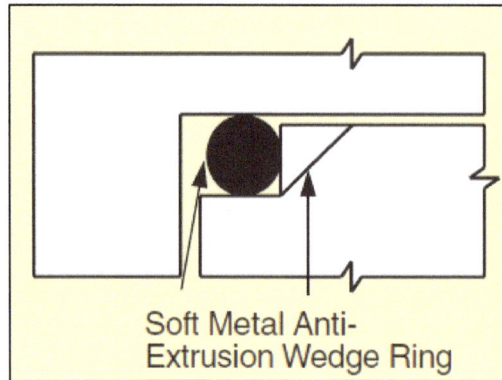

Soft Metal Anti-Extrusion Wedge Ring

Fig. 3.10 - Seal Extrusion Avoidance by use of Wedge-Rings (Courtesy of Parker)

3.3.4- Design Issues - Gland (Groove) Sharp Corners

Failure Source: Improper groove design with sharp corners.

Failure Mode: A seal groove is not adequately rounded. As shown in Fig. 3.11, an O-Ring was used in a stack valve at 250 bar (3,625 psi) was damaged at the outer circumference even with relatively small seal gaps.

Suggested Solution: Design seal gland (groove) to meet the design codes.

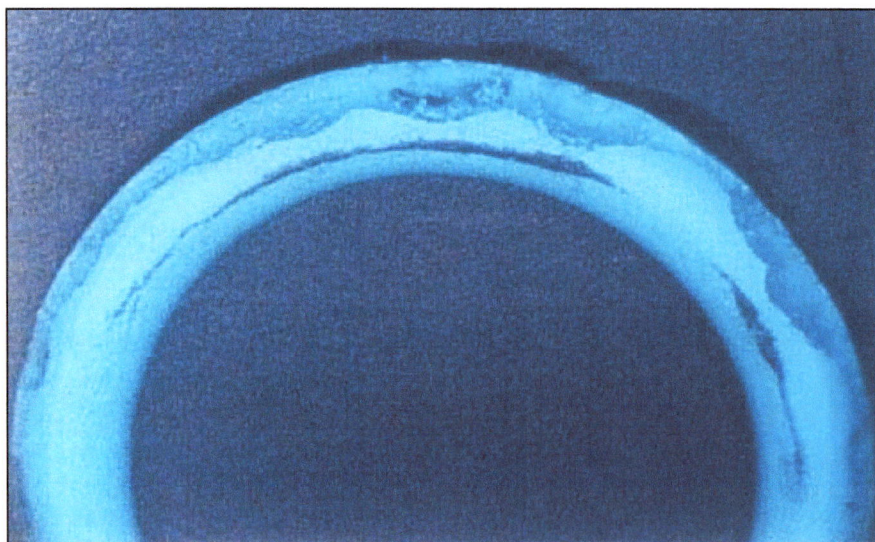

Fig. 3.11 - O-Ring Failure due to Sharp Corners and High Pressure (www.o-ring-lab.com)

3.3.5- Design Issues - Rough Surfaces

Failure Source: Finishes of contact surfaces have much to do with the life of dynamic seals. A seal can be exposed to abrasion due to:
- Contact with very rough surface results in high friction with the seal material.
- Contact with very smooth surface results in running the seal dry without lubrication.

Failure Mode: Figures 3.12 through 3.15 show hydraulic seals abrasion marks.

Suggested Solution: The surface must be rough enough to hold small amounts of oil for lubrication. The most desirable surface finish value is from 10 to 20 micro-inches. However, surface finish values less than 5 micro-inches are not recommended for dynamic seals. Otherwise, an extending cylinder rod will be wiped completely dry and will not be lubricated when it retracts.

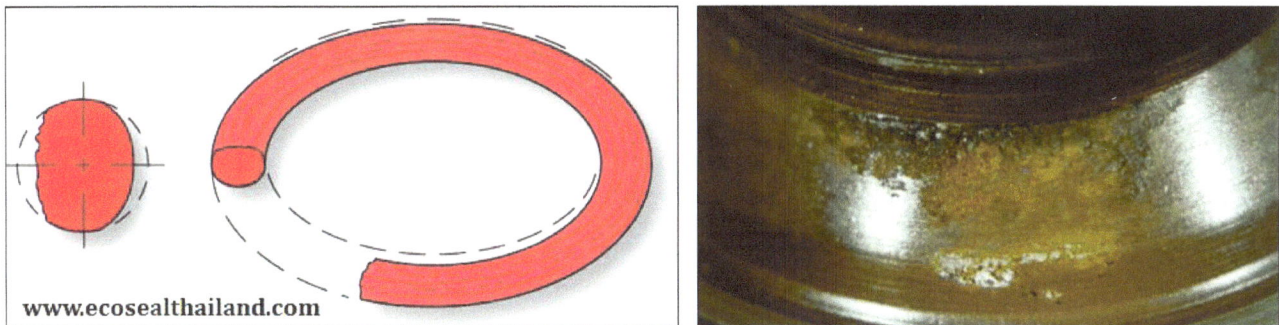

www.ecosealthailand.com

Fig. 3.12 - Seal Abrasion due to Contact with Rough Surface, Example 1

NBR piston u-cup with abrasion marks at the seal lip and migrating across the dynamic surface

**Fig. 3.13 - Seal Abrasion due to Contact with Rough Surface, Example 2
(Courtesy of System Seals Inc.)**

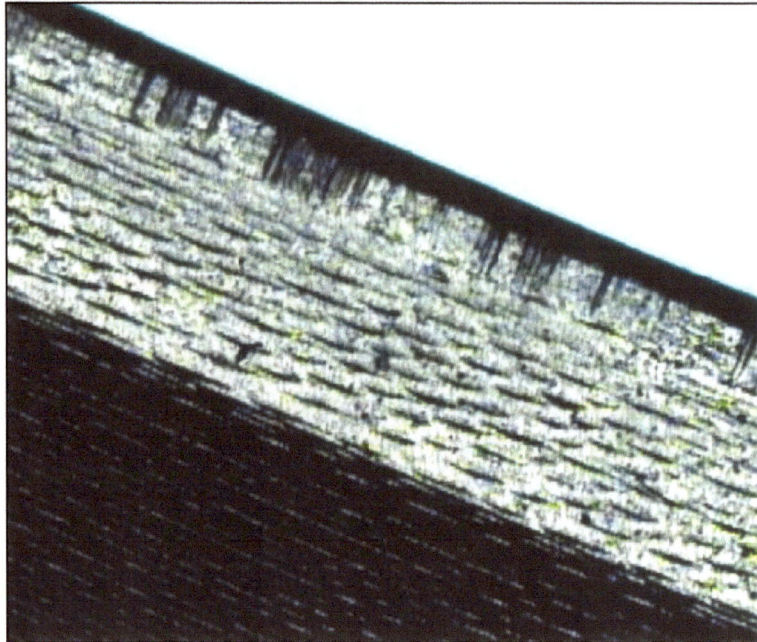

U-cup seal lip showing abrasion marks

**Fig. 3.14 - Seal Abrasion due to Contact with Rough Surface, Example 3
(Courtesy of System Seals Inc.)**

PTFE seal with abrasion marks at the seal lip and migrating across the dynamic surface.

**Fig. 3.15 - Seal Abrasion due to Contact with Rough Surface, Example 4
(Courtesy of System Seals Inc.)**

3.3.6- Design Issues - Blow-By Effect

Failure Source: Figure 3.16 shows a traditional design of a bidirectional piston sealing solution. In such design, *Blow-By Effect* occurs causing leakage.

Failure Mode: Increased leakage rate.

Suggested Solution: As shown in Fig. 3.17, to resolve this problem, piston seal should contain radial grooves or notches. One notch for a unidirectional seal or two notches for a bidirectional seal. With such notches, fluid pressure reaches out to the O-Ring and compresses it in a way that controls dynamically the clearance between the piston seal and the cylinder wall. As a result, leakage across the piston is controlled.

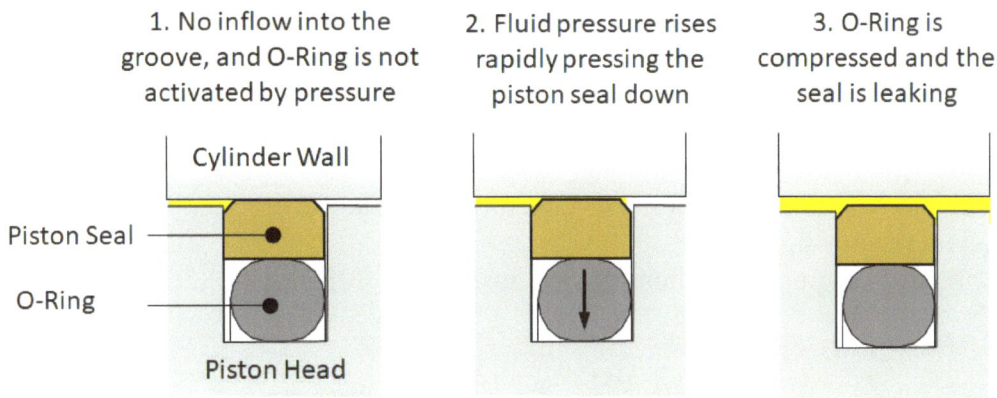

Fig. 3.16 - Blow-By Effect

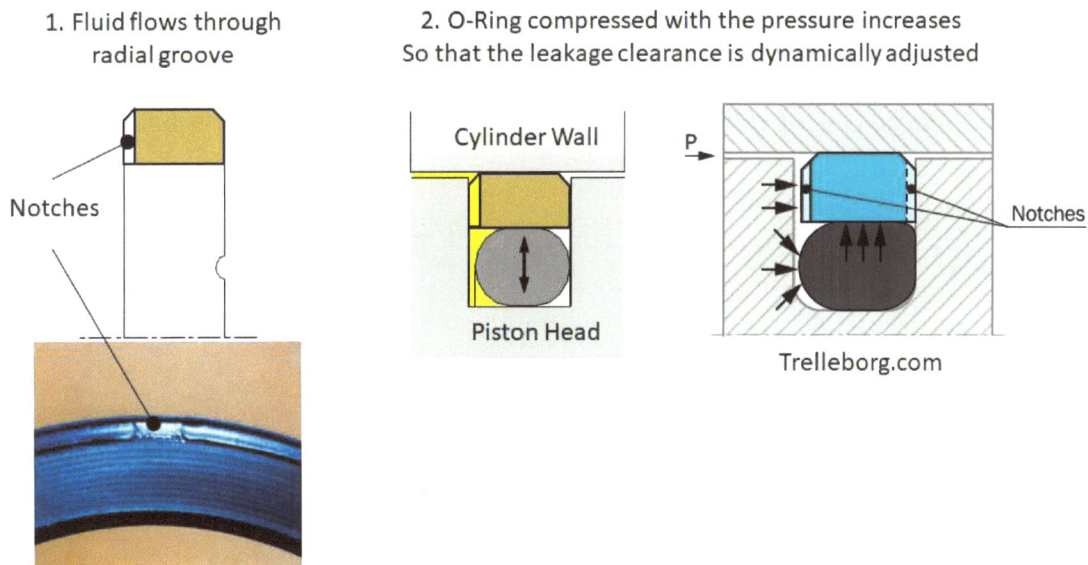

Fig. 3.17 - Resolving Blow-By Effect

3.3.7- Assembly Procedures - Passing Over Sharp Edges

Failure Source: Damage to hydraulic seals can occur during installation when:
- Seals come in contact with sharp edges such as threads.
- Insufficient lead-in chamfer.
- Oversize piston seal.
- Undersize rod seal.
- Seal is twisted/pinched during installation.
- Seal is not properly lubricated before installation.
- Seal is dirty or contaminated with metal particles upon installation.

Failure Mode: As shown in Fig. 3.18, failure can occur due to passing over sharp edges in the gland area. As shown in Fig. 3.19, short cuts can happen due to passing over sharp edges in the lead-in chamfer or sharp threads.

Suggested Solution:
- Cover threads and sharp edges before assembly.
- Use of proper installation tools.
- Make sure lead in chamfers are based on manufacturers recommendations.
- Select proper seal size.
- Consider proper cleanliness during assembly.

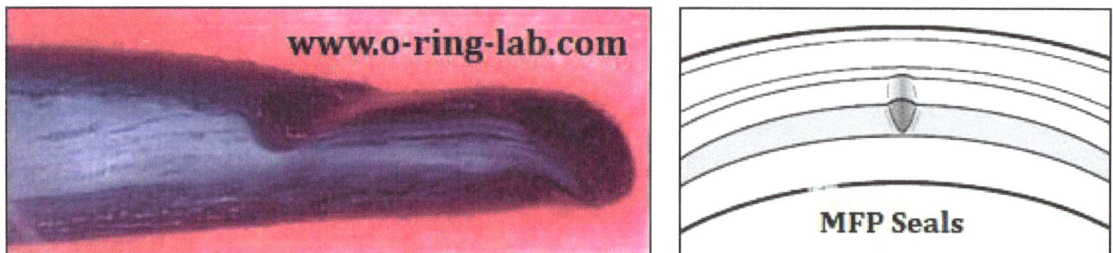

Fig. 3.18 - Sheared O-Ring due to Passing Over Sharp Edges in the Gland Area During Assembly (www.o-ring-lab.com)

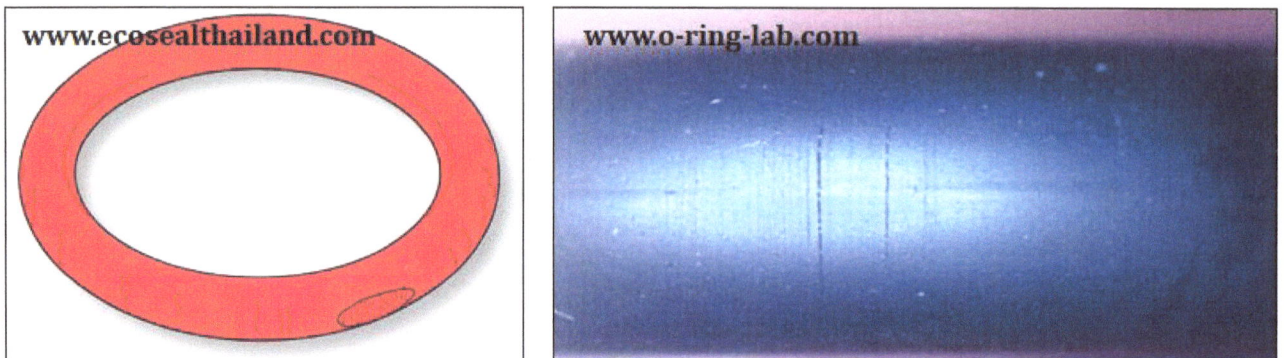

Fig. 3.19 - O-Ring Failure due to Passing Over Guide Chamfer

3.3.8- Operational Conditions - Overpressure

Failure Source: overpressure work environment.

Failure Mode: As shown in Figures 3.20 and 3.21, hydraulic seals are stressed beyond their limits and fail.

Suggested Solution:
- Work within recommended pressure range.
- Select seal material based on maximum working pressure

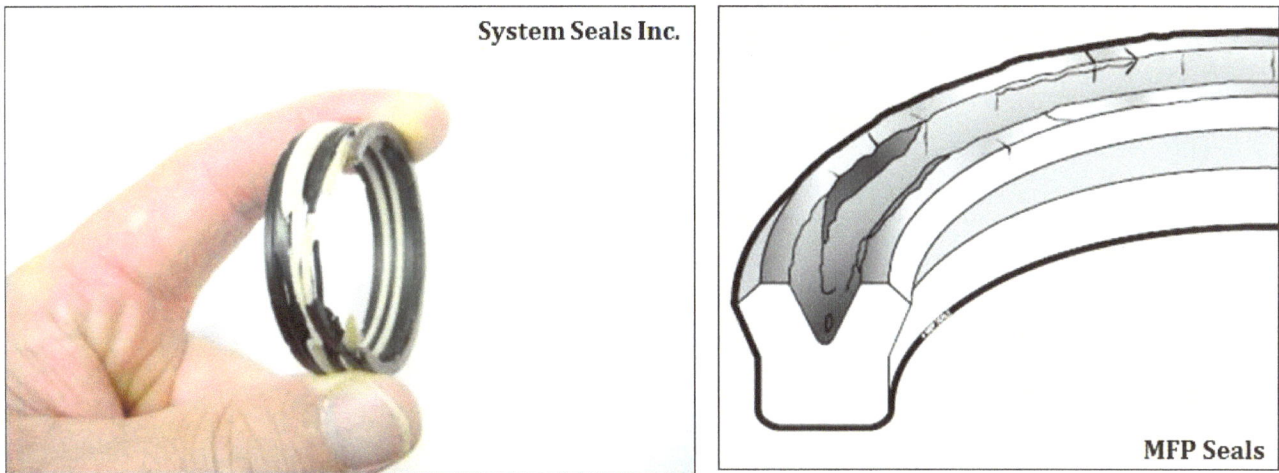

V-packing set that structurally failed at high pressure

Fig. 3.20- Seal Failure due Overpressure, Example 1

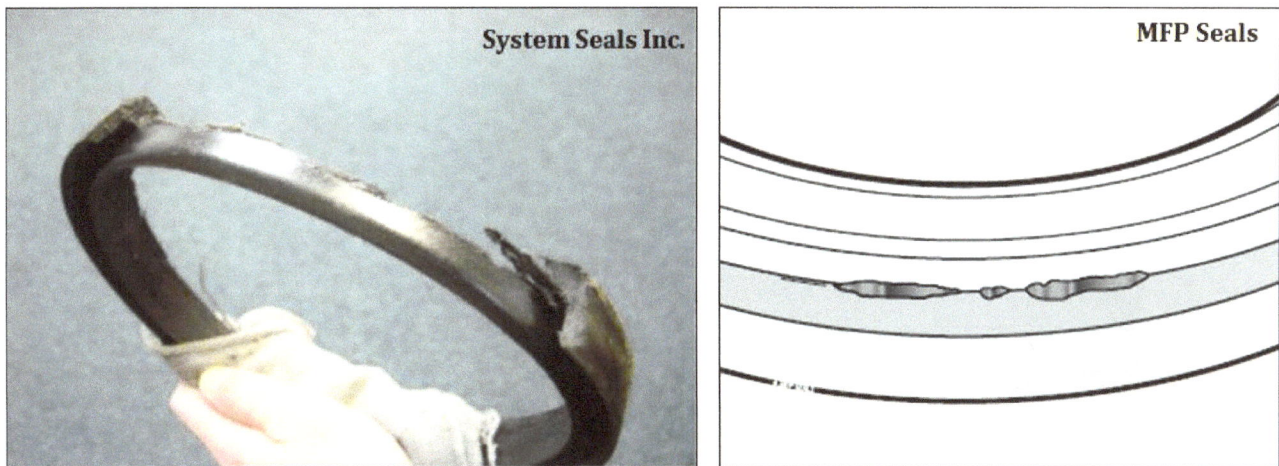

Rubber and fabric piston seal that failed from over-pressurization

Fig. 3.21 - Seal Failure due Overpressure, Example 2

3.3.9- Operational Conditions - Pressure Trapping

Failure Source: Pressure spikes, such as those created by sudden shifting of a directional valve, may be ten times greater than the normal operating pressure of a hydraulic seal. If pressure spikes occur often, as shown in Fig. 3.22, pressure is trapped between seals causing seal damage.

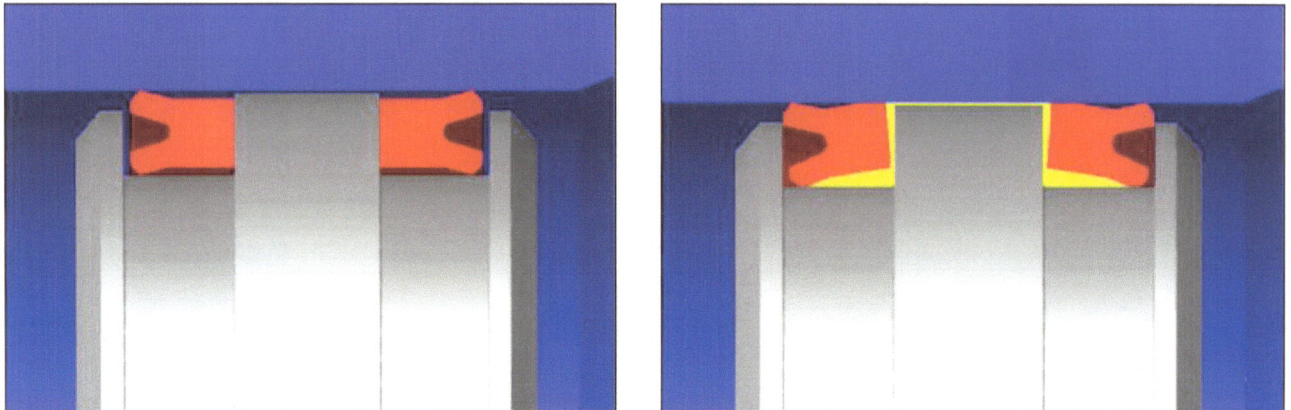

Normal action of piston can cause certain types of piston seals to trap pressure. Excessive pressure between the seals can push the seals away from each other, ultimately resulting in pressure trapping failure

Fig. 3.22 - Bidirectional Seal Failure due to Overpressure (Courtesy of System Seals Inc.)

Failure Mode: As shown in Figures 3.23, a hydraulic seal is squeezed and failed as a result of *Pressure Trapping*.

Suggested Solution: Apply design strategies to eliminate or at least minimize creating pressure shocks.

Left: Back-to-back piston u-cup seals showing pressure trapping with reverse extrusion as a result
Right: Loaded polyurethane u-cups showing severe pressure trapping and failure.

Fig. 3.23 - Seal Failure due Pressure Trapping (Courtesy of System Seals Inc.)

3.3.10- Operational Conditions - Overheating

Failure Source:
- Long exposure to high working temperature or excessive heat.
- High speed operation that increases the seal lip temperature.

Failure Mode: Heat generated by friction or directly through other heat sources accelerates hardening of the seal material, particularly in the contact area between the sealing lip and the sliding surface. This leads to cracks which become increasingly larger over time and ultimately result in seal failure. Figures 3.24 through 3.27 shows various failure modes such as: hardening and cracks of the sealing lip, softening permanent deformation of the seal body, extrusion, splits, ruptures, melting, and squeezing. It is to be noted also that ow temperature, beyond the seal material's specified minimum temperature, makes the seal brittle.

Suggested Solution:
- Work within recommended temperature range.
- Select proper seal material that works better at high temperature.

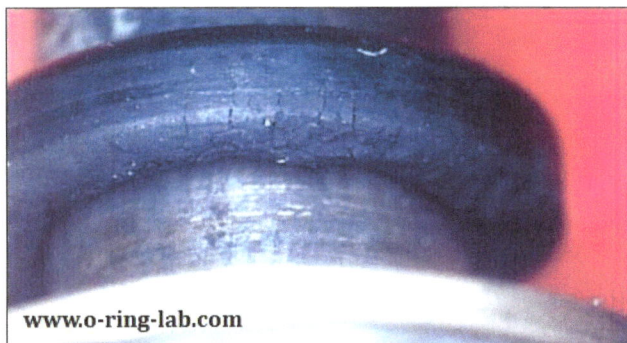

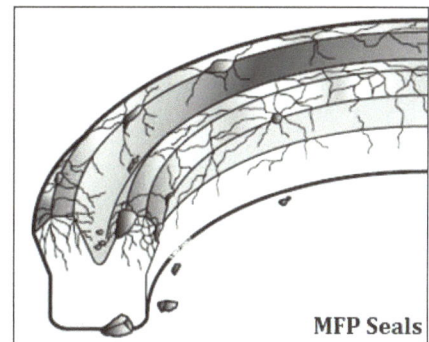

Fig. 3.24 - Seals Hardened and Cracked due to Long-Term Exposure to High Temperature

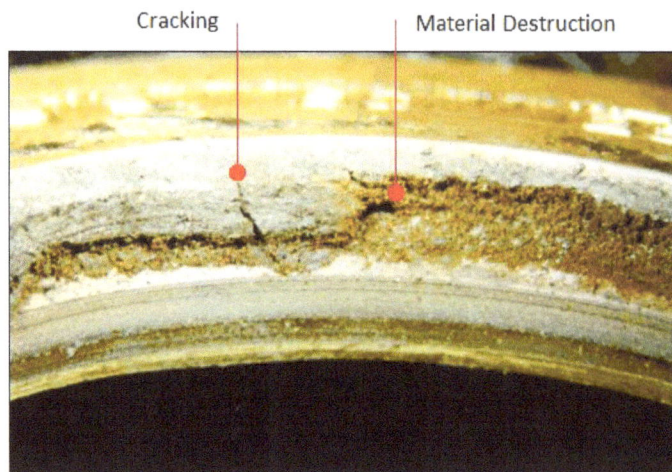

Fig. 3.25 - Shaft Seal Failure due to Exposure to Excessive Working Temperature with Grease Lubricant

Fig. 3.26 - V-Packing Failure due to Exposure to Excessive Working Temperature

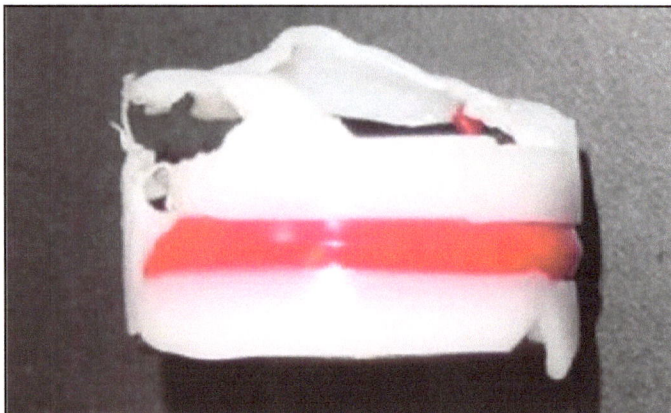

Fig. 3.27 - Multi-Component Piston Seal Melted due to Exposure to Excessive Working Temperature (Courtesy of System Seals Inc.)

3.3.11- Operational Conditions – Over-Speeding

Failure Source: High surface speed generates excessive heat.

Failure Mode: As shown in Fig 3.28, the dynamic seal is hardened showing cracks and glazing of the seal material.

Suggested Solution:
- Reduce surface speed (stroke speed or RPM).
- Select proper seal material that works better at high speed and high temperature.

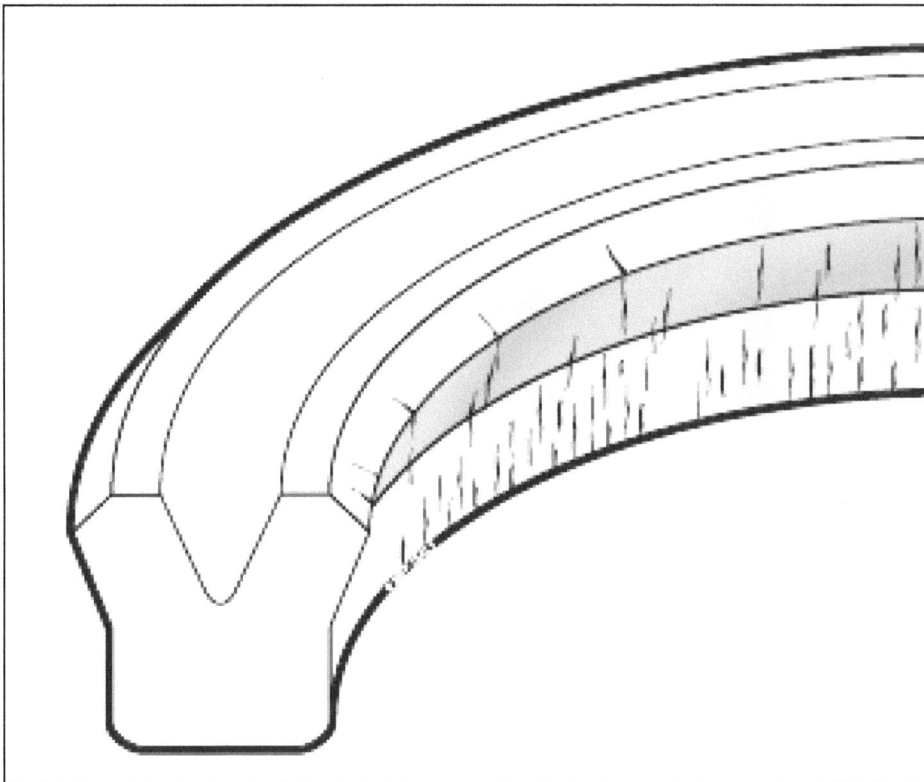

**Fig. 3.28 - Hardened and Cracked Dynamic Sealing Surface due to Overspeeding
(Courtesy of MFP Seals)**

3.3.12- Operational Conditions - Contamination

Failure Source:
- Hydraulic fluids contaminated by abrasive and metal particles.
- Dirty assembly area.
- Poor wiper performance.
- Highly contaminated work environment around a cylinder as shown in Fig. A.162.

Failure Mode: Figures 3.29 and 3.31 show damaged hydraulic seals due to contamination.

Suggested Solution: Make sure hydraulic fluids comply with the cleanliness level recommended by the system manufacturer.

Fig. 3.29 - Hydraulic Cylinder Operating in a Severe Salt Contaminated Work Environment (Courtesy of System Seals Inc.)

The metal particles embedded in the seal produce scores on the mating surface.

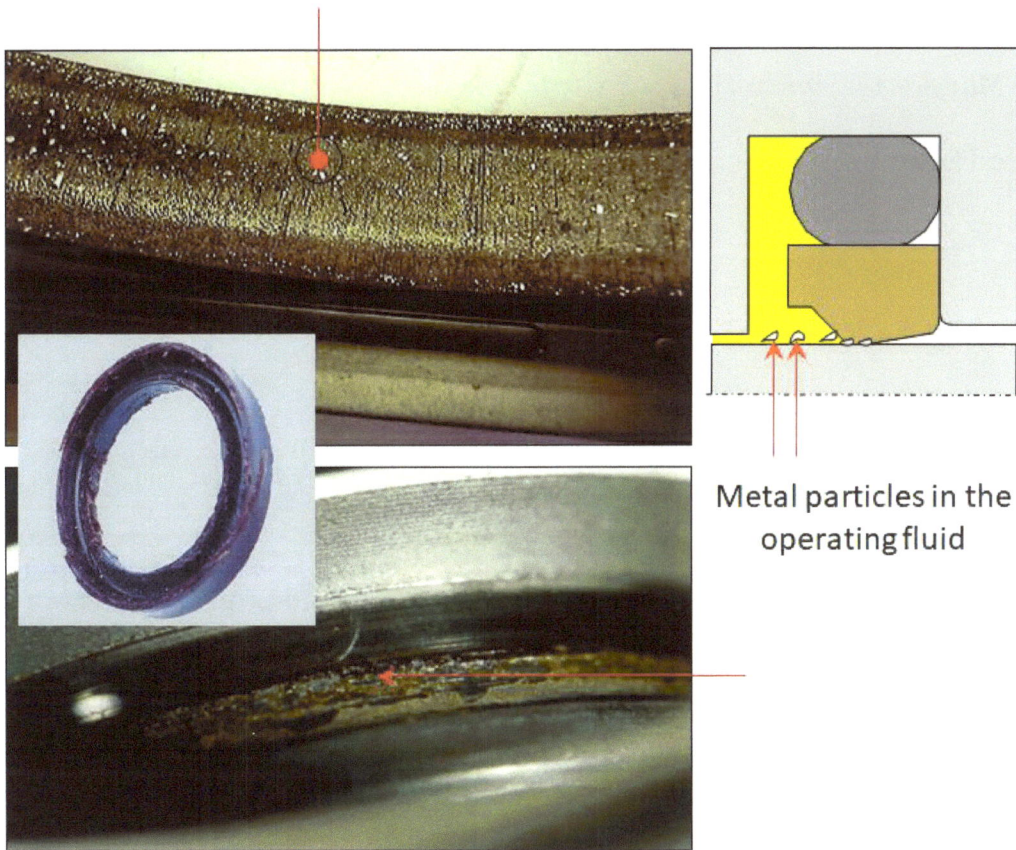

Metal particles in the operating fluid

Fig. 3.30 - Seal Failure due to Contamination

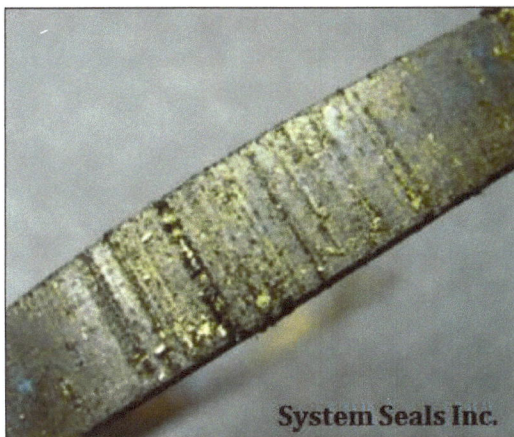

System Seals Inc.

MFP Seals

Resin/Fabric guide band damaged by severe metallic contamination

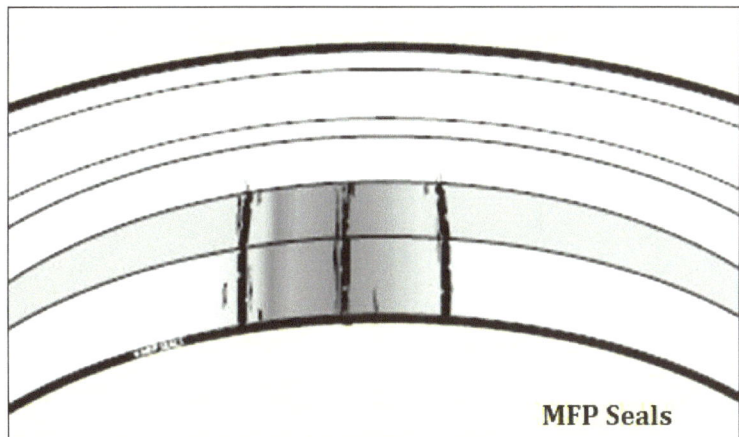

Dynamic seal lip shows axial cuts and grooves

Fig. 3.31 - Seal Failure due to Contamination

3.3.13- Operational Conditions - Fluid Incompatibility

Failure Source: incompatibility with the operating hydraulic fluid.

Failure Mode: As shown in Fig. 3.32, a hydraulic seal lost its flexibility and cracks are formed.

Suggested Solution: Make sure hydraulic seals and hydraulic fluids are compatible.

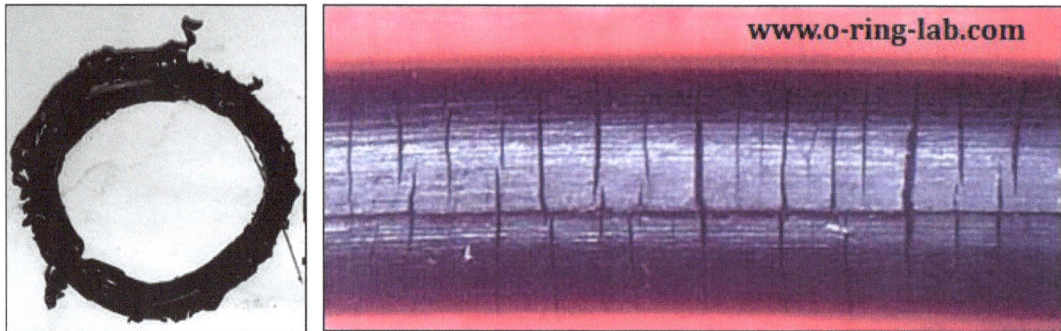

Fig. 3.32 - Seal Failure due to Hydraulic Fluid Incompatibility (www.o-ring-lab.com)

3.3.14- Operational Conditions - Chemical Attack

Failure Source: Chemical interaction between the seal and the hydraulic fluid.

Failure Mode: As shown in Fig. 3.33, a hydraulic seal was subjected to unfavorable defects such as excessive hardening, softening, swelling, and shrinkage.

Suggested Solution:
- Make sure acidity of the hydraulic fluid is within allowable limits.
- Check the resistivity level of the hydraulic seals to chemical attacks.

Fig. 3.33 - Seal Failure due to Hydraulic Fluid Chemical Attack

3.3.15- Operational Conditions - Hydrolysis

Failure Source: exposure to water or water-based fluids at elevated temperatures.

Failure Mode: Figures 3.34 and 3.35, show break-down of the seal material, loss of physical properties, cracking, and crumbling of the material due to *Hydrolysis* Failure.

Suggested Solution: Select proper seal material for water-based fluids.

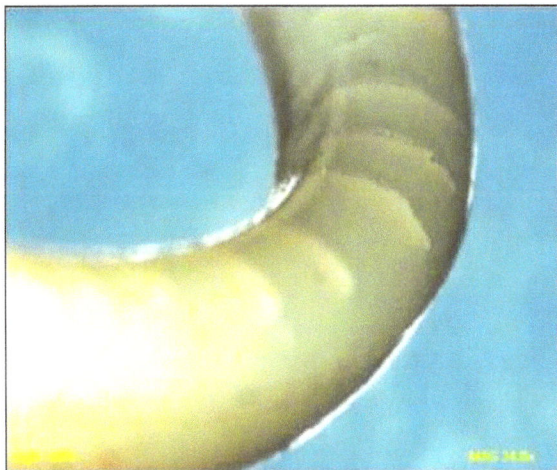

Polyurethane O-Ring

Fluorocarbon (FPM) seal

Fig. 3.34 - Seals Showing Early Signs of Hydrolysis (Courtesy of System Seals Inc.)

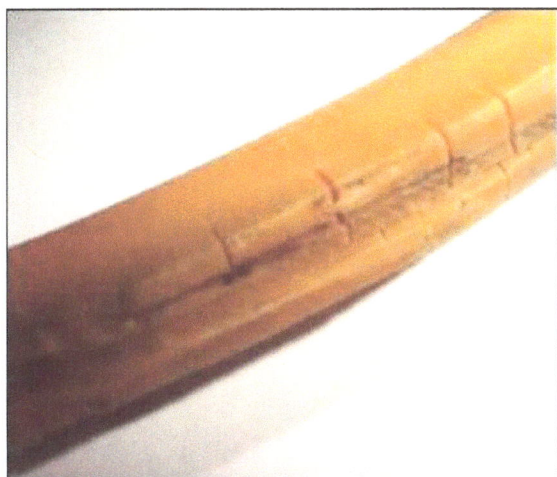

Thermoplastic elastomer seal

Fluorocarbon (FPM) seal

Fig. 3.35 - Seals Showing Late Signs of Hydrolysis (Courtesy of System Seals Inc.)

3.3.16- Operational Conditions – Explosive Decompression

Failure Source: Sudden decompression of air bubbles. Mineral oils contain, at atmospheric pressure, up to 10% by volume molecularly dissolved air. In a "saturated" condition, the dissolved air has no effect on the hydraulic oil performance. If the pressure on the oil falls, the high volume of air molecules can no longer remain in solution, and the air will separate and form bubbles. When the operating fluid is aerated, air bubbles expand in an explosive manner very quickly upon sudden pressure drops resulting in damage to the sealing elements.

Failure Mode: As shown in Figures 3.36 through 3.38, hydraulic seals failed due to air bubbles sudden decompression.

Standard Test Method (Explosive Decompression Test): A high-pressure test rig is used to pressurize and depressurize the sealing element at a certain frequency under specified temperature.

Suggested Solution:
- Apply the required design strategies to prevent hydraulic fluid aeration.
- Apply design strategies to control the decompression rate of pressurized fluid.
- Use *Anti-Explosive Decompression* (AED) seals.

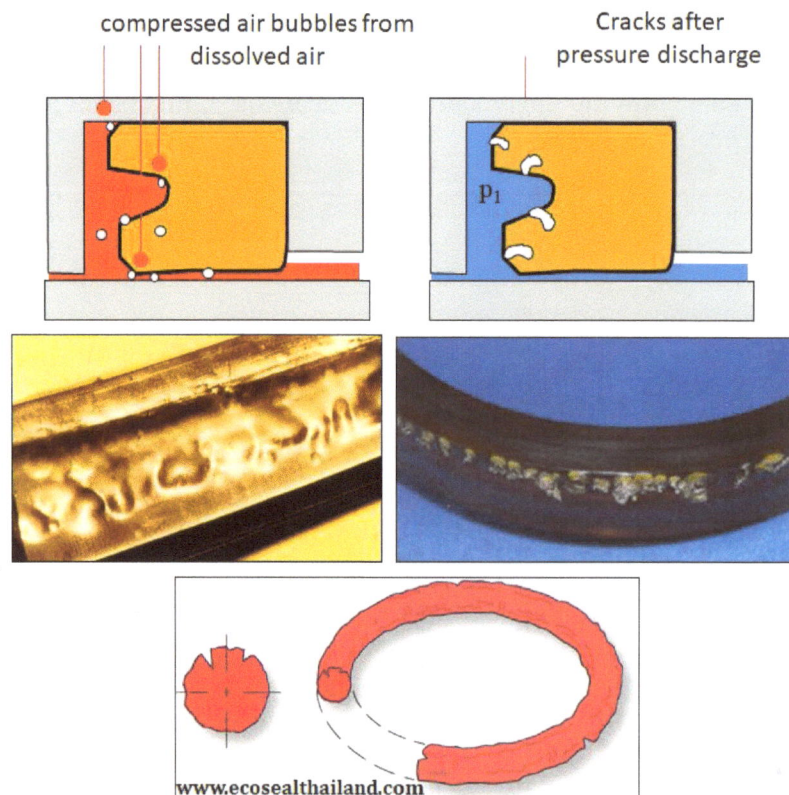

Fig. 3.36 - Seal Failure due to Explosive Decompression, Example 1

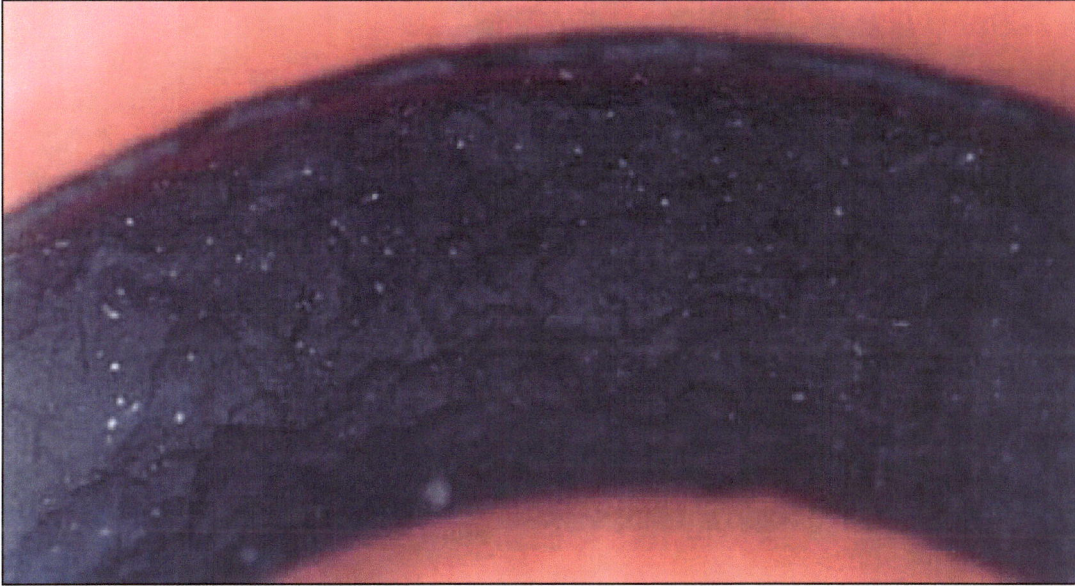

**Fig. 3.37 - Seal Failure due to Explosive Decompression, Example 2
(www.o-ring-lab.com),**

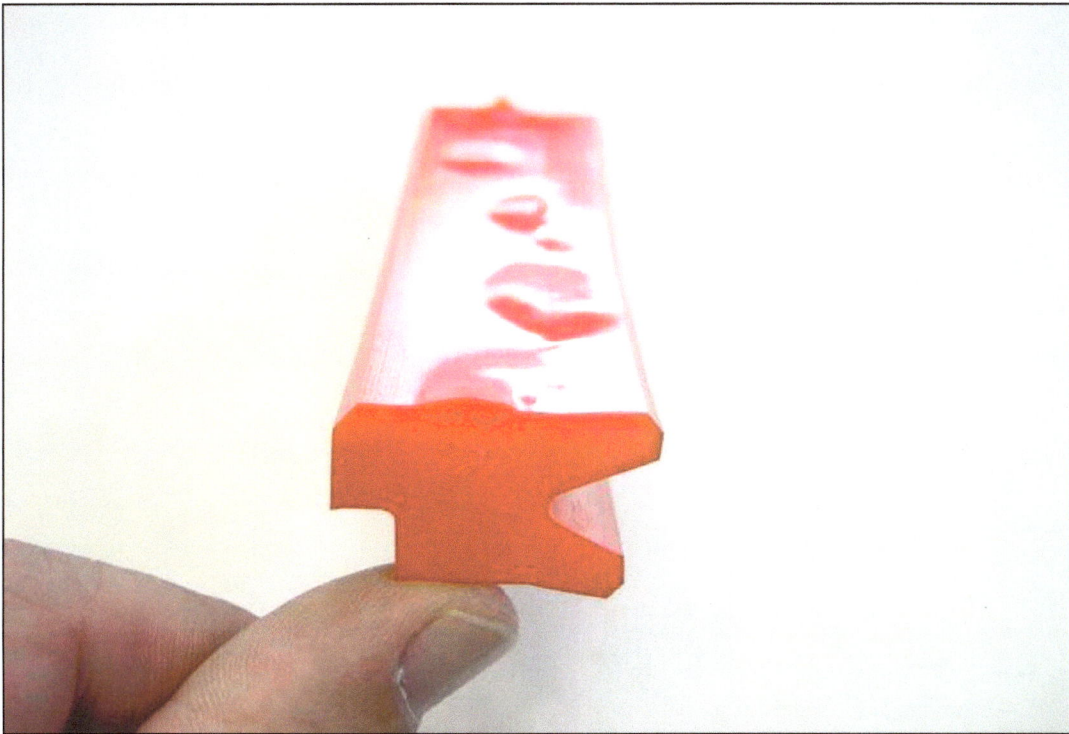

Polyurethane u-cup showing severe dieseling damage

**Fig. 3.38 - Seal Failure due to Explosive Decompression, Example 3
(Courtesy of System Seals Inc.)**

3.3.17- Operational Conditions - Dieseling

Failure Source: Sudden compression of air bubbles. When the operating fluid is aerated, air bubbles are compressed and self-ignite very quickly upon sudden pressure increase with the presence of extreme working temperature. This is known as the *Dieseling Effect* like fuel burning in diesel engines.

Failure Mode: As shown in Figures 3.39 and 3.40, hydraulic seals are burned and damaged as a result of dieseling effect.

Suggested Solution:
- Apply the required design strategies to prevent hydraulic fluid aeration.
- Apply design strategies to control the working temperature and pressure shocks.

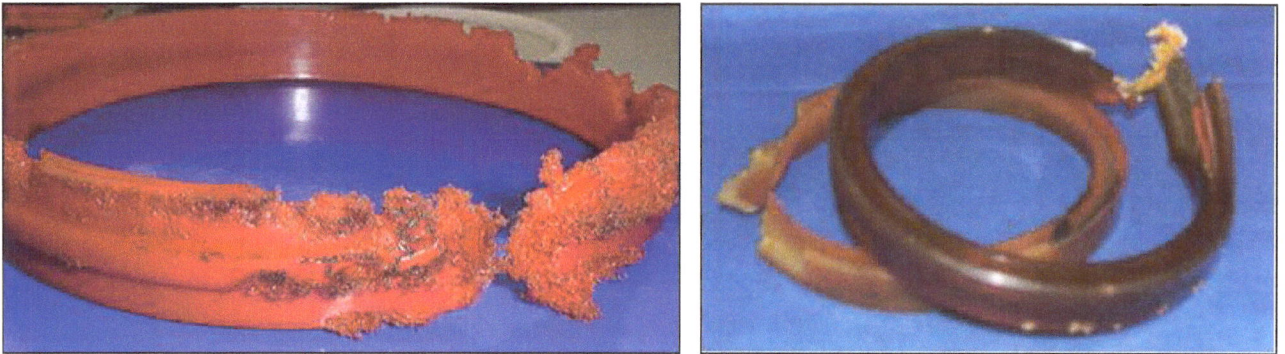

Fig. 3.39 - Seal Damage due to Diesel Effect, Example 1

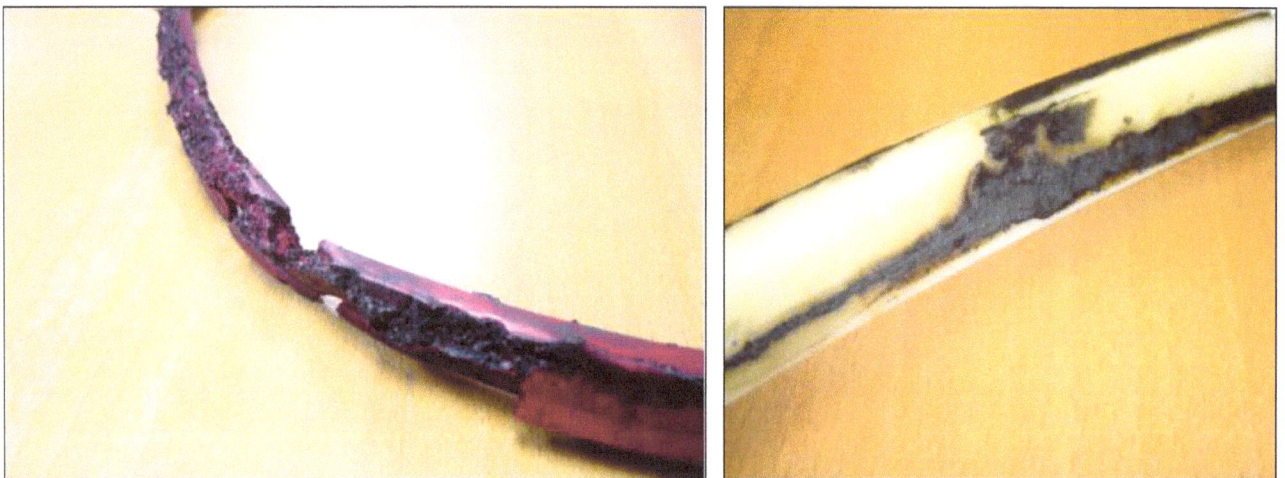

Polyurethane u-cup showing severe dieseling damage   *Nylon Back-up ring with dieseling damage*

Fig. 3.40 - Seal Damage due to Diesel Effect, Example 2 (Courtesy of System Seals Inc.)

3.3.18- Operational Conditions - Side Loading

Failure Source:
- Side loads on a cylinder piston or cylinder rod.
- Insufficient load guidance.

Failure Mode: Figures 3.41 shows damage to the sealing elements and the cylinder components as a result of side loading. The following are additional consequences of side loading:
- Increased gland clearance on one side only.
- Gap extrusion.
- Increased leakage.
- Uneven friction on the seal.
- Rod or barrel will be galled or scored.

Suggested Solution:
- Follow the best practices for mounting the hydraulic cylinder and the attached load.
- Whenever possible, use *Stop Tubes* to reduce the effect of side loads.
- Proper seal package design that includes Guide-Rings.

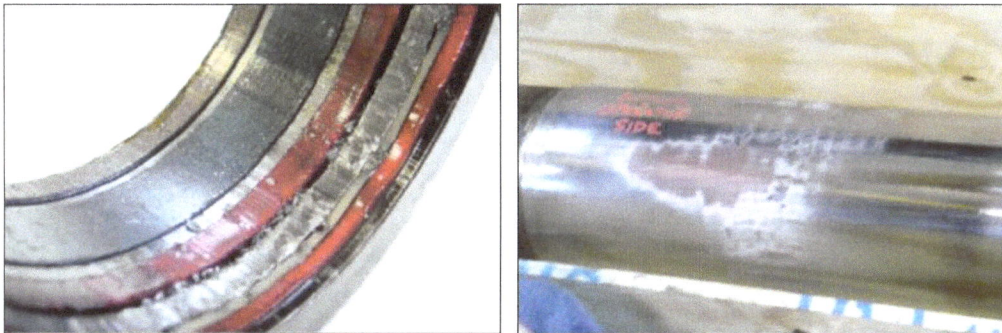

Left: Severe side loading with catastrophic damage from metal-to-metal contact
Right: Side loading that wore the chrome off the rod

Fig. 3.41 - Seal Damage due to Side Loading (Courtesy of System Seals Inc.)

3.3.19- Operational Conditions - Vibration

Failure Source: small frequent motions which are usually encountered when equipment is in transit. Such defects are reported in hydraulic cylinders more than any other components.

Failure Mode: Excessive wear of hydraulic seals.

Suggested Solution: Apply the required design strategies to isolate the vibrations from hydraulic components.

3.3.20- Operational Conditions - Spiral Failure

Failure Source: This failure occurs when some segments of the O-ring slide while other segments simultaneously roll. The design and operational factors which contribute to spiral failure of a seal are listed below in the order of their relative importance:

1. Speed of stroke.
2. Lack of lubrication.
4. Squeeze and softness of sealing rings.
5. too much space for movement in the groove.
6. Temperature of operation.
7. Length of stroke.
8. Surface finish of gland.
9. Type of metal surface.
10. Side loads.
11. ID to W ratio of O-ring.
12. Contamination or gummy deposits on metal surface.
15. Eccentricity of sealing ring.
16. Stretch and softness of sealing rings.
17. Lack of Back-up Rings.

Failure Mode: A unique type of failure, called *Torsional or Spiral Failure*, may occur on reciprocating dynamic sealing rings of different cross section. This failure was given this name because when it occurs, as shown in Fig. 3.42, the seal has spiral 45-degree angle deep cuts through the crosssection in a spiral pattern.

Suggested Solution: Resolve the previously mentioned sources of such a failure.

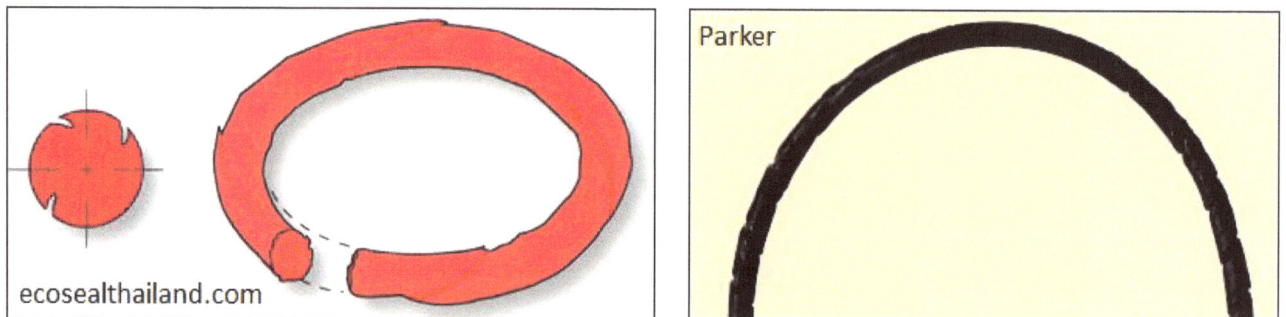

Fig. 3.42 - Sealing Ring Spiral Failure (ecosealthailand.com)

3.3.21- Operational Conditions - Seal Wear

Failure Source: The wear pattern should be even and consistent around the circumference of the dynamic lip. A small amount of even wear will not drastically affect seal performance; however, if the wear patterns are uneven or grooved, or if the amount of wear is excessive, performance may be dramatically reduced. Table 3.3 lists the factors that influence seal wear.

Factors that Influence Seal Wear	
Rough surface finish	Excessive abrasion may occur
Ultra-smooth surface finish	Surface finishes below 2 µm Ra can create aggressive seal wear due to lack of lubrication
High pressure	Increases the radial force of the seal against the dynamic surface
High temperature	While hot, materials soften, thus reducing tensile strength
Poor fluid lubricity	Increases friction and temperature at sealing contact point
Tensile strength of seal compound	Higher tensile strength increases the material's resistance to tearing and abrading
Fluid incompatibility	Softening of seal compound leads to reduced tensile strength
Coefficient of friction of seal compound	Higher coefficient materials generate higher frictional forces
Abrasive fluid or contamination	Creates grooves in the lip, scores the sealing surface and forms leak paths
Extremely hard sealing surface	Sharp peaks on hard surfaces will not be rounded off during normal contact with the wear rings and seals, accelerating wear conditions

Table 3.3 - Factors Affecting Seal Wear (Courtesy of Parker)

Failure Mode: As shown in Fig. 3.43, uneven, grooved, or excessive wear in dynamic seals.

Suggested Solution: Resolve the previously mentioned sources of such a failure.

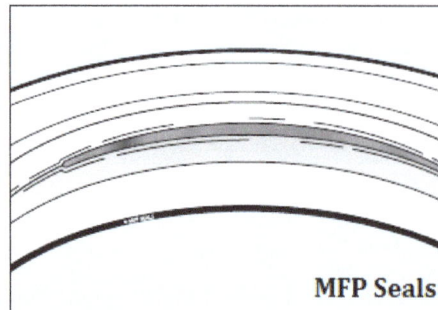

Only one side of the dynamic lip is showing excessive wear.

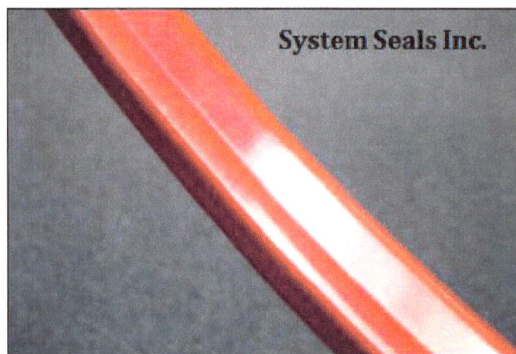

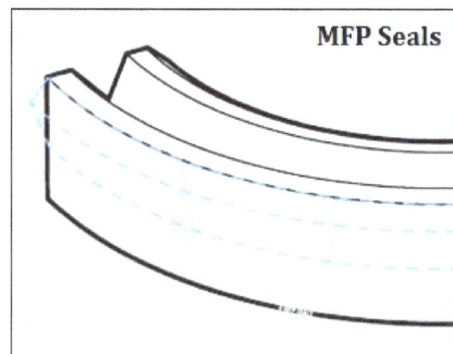

Left: Polyurethane Rod Seal with Shiny and Smooth Surface from a Dry Running Condition
Right: The dynamic face of the seal is worn to a glossy mirror like shine.

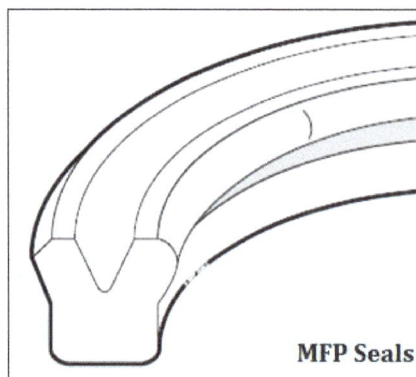

The dynamic lip is worn to a rounded, egg-shape.

Fig. 3.43 - Failure Modes of Hydraulic Seals due to Wear (Courtesy of MFP Seals)

3.3.22- Operational Conditions - Fatigue

Failure Source: Cold startup and/or Exposure to cyclic pressure with high frequency.

Failure Mode: Figure 3.44 shows that the V portion of the seal shows long cracks or splits.

Suggested Solution: proper selection of seal design and material. Control the startup temperature.

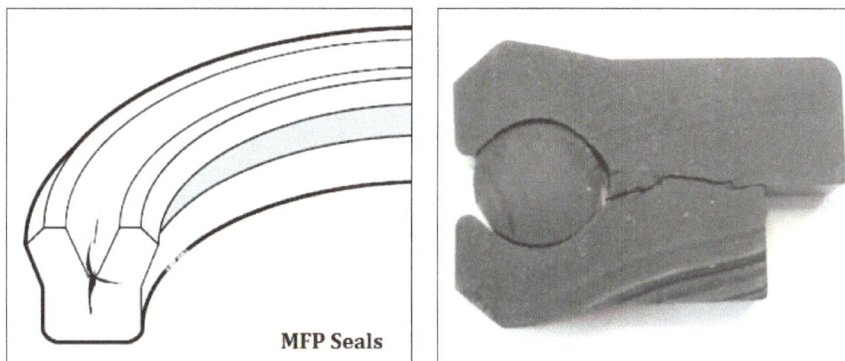

MFP Seals

**Fig. 3.44 - Profile Section of an O-Ring Loaded U-Cup with Flex Fatigue Cracking
(Courtesy of System Seals Inc.)**

3.3.23- Normal Aging - Hardening

Failure Source: Normal *Aging*.

Failure Mode: Figure 3.45 shows cracks on the inner circumference of the sealing ring.

Suggested Solution: Use the seals within their estimated lifetime. Make sure storage is arranged based of first-come first-use basis.

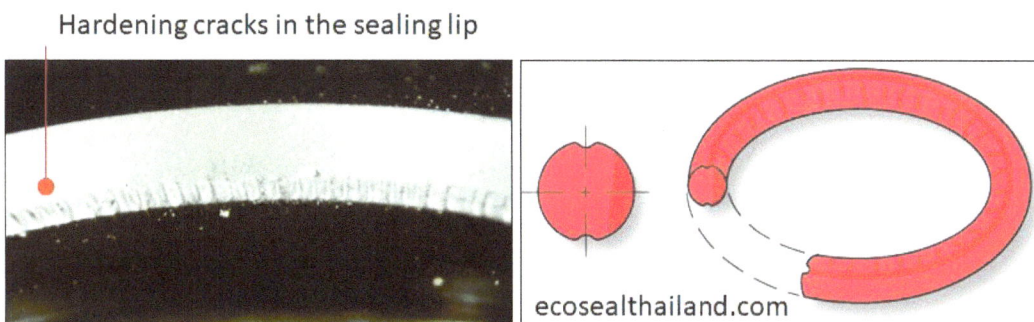

Hardening cracks in the sealing lip

ecosealthailand.com

Fig. 3.45 - Sealing Ring Hardening due to Normal Aging

3.3.24- Normal Aging - Splits

Failure Source: Expired *Shelf Life.*

Failure Mode: Figure 3.46 material split through the seal body.

Suggested Solution: Use the seals within their estimated lifetime.

Fig. 3.46 - Failure due to Normal Aging (www.o-ring-lab.com)

3.3.25- Storage Conditions - Swelling

Failure Source: Elastomers have a higher coefficient of thermal expansion than steel. This means the seal will expand more when it is hot. Same problem could happen due to fluid incompatibility, chemical attack, and when the seal absorbs moisture.

Failure Mode: Figure 3.47 shows a seal that absorbed the surrounding water like a sponge and swells to the point of malfunction. Figure 3.48 shows a case of wiper swelling.

Suggested Solution:
  ▪ Adjust the humidity level at the storage space.
  ▪ Test the volume change of hydraulic seals periodically.

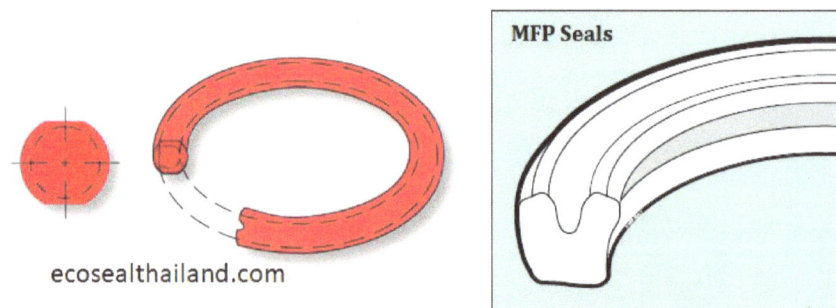

Fig. 3.47 - Sealing Ring Excessively Swells due to Humid Storage Space (ecosealthailand.com)

Fig. 3.48 - Swelling Failure of a wiper (Courtesy of System Seals Inc.)

3.3.26- Storage Conditions - Ozone Cracking

Failure Source: This damage is a result of exposure of a sealing ring to Ozone for several weeks without protection.

Failure Mode: Figure 3.49 shows many small surface cracks perpendicular to the direction of stress.

Suggested Solution: Provide proper protection against Ozone.

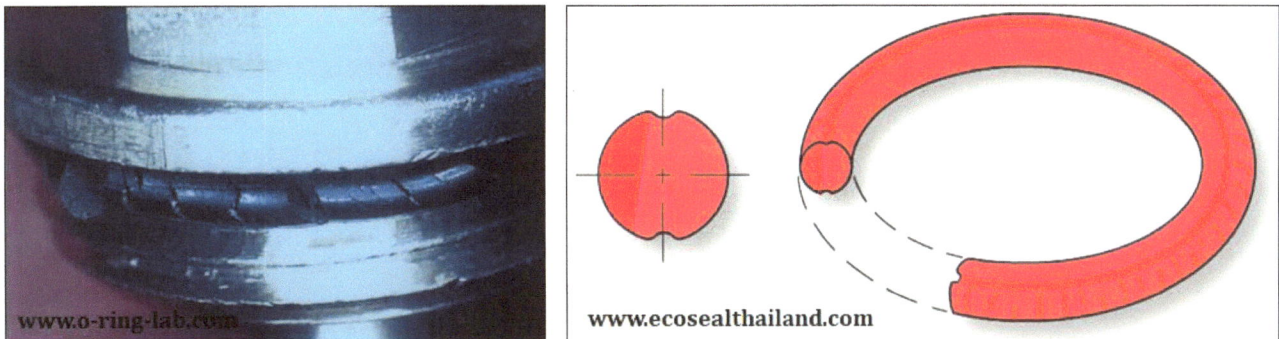

Fig. 3.49 - Sealing Ring Surface Cracking due to Exposure to Ozone (ecosealthailand.com)

Chapter 4
Troubleshooting and Failure Analysis of Pumps

Objectives

This chapter discusses hydraulic *pumps* inspection, troubleshooting, and failure analysis. In this chapter, troubleshooting charts for twelve different faults of hydraulic pumps are presented. The chapter also presents examples of defective pumps due to contamination, overheating, cavitation, and fatigue stress for gear, vane, and piston pumps.

Brief Contents

4.1- Hydraulic Pumps Inspection

4.2- Hydraulic Pumps Troubleshooting

4.3- Hydraulic Pumps Failure Analysis

Chapter 4: Troubleshooting and Failure Analysis of Pumps

4.1- Hydraulic Pumps Inspection

Hydraulic *pumps* have various characteristics based on pump displacement (fixed or variable), rotation (unidirectional, bidirectional, and over center), and pumping mechanism (gear, vane, and piston). Volume 1 of this series of textbooks presents an overview about the construction and operating principle of various pump mechanisms. Table 4.1 shows a typical inspection sheet for a pump.

Hydraulic Pump Inspection Sheet	
Manufacturer	
Model #	
Serial #	
Location	
Pumping Mechanism	☐ External Gear ☐ Internal Gear ☐ Gerotor ☐ Vane Pump [☐ Balanced ☐ Unbalanced] ☐ Radial Piston [☐ Rotating Cam ☐ Rotating Cylinder Block] ☐ Bent Axis ☐ Swash Plate ☐ Other [ ]
Direction of Rotation	☐ Unidirectional ☐ Bidirectional ☐ Over Center
Pump Displacement	☐ Fixed ☐ Variable [= cc/rev]
Type of Control	☐ Pressure Compensated ☐ Displacement Controlled ☐ Constant Power (Torque) ☐ Load Sense
Drive Shaft	Type and Size:
Ports	Case Drain: ☐ Yes ☐ NO Case Drain size: Inlet Port size: Outlet Port Size:
Conditions of Seals	
Conditions of Bearings	
Conditions of Inside Parts	
Other Nots	

Table 4.1 – Hydraulic Pumps Inspection Sheet

4.2- Hydraulic Pumps Troubleshooting

4.2.1- No Flow Out of the Pump

Table 4.2 shows the troubleshooting actions when no *flow* out of the pump is observed.

T-Pump-01-No Flow out of the Pump	
Pump drive motor is not working.	▪ Check wiring connection to electric motor. ▪ Check wiring connection in control circuit. ▪ Check use of correct voltage. ▪ Check fuses in the electrical control circuit.
Pump rotates in wrong direction.	▪ Check polarity of electrical motor.
Pump is not receiving fluid.	▪ Check oil level in the reservoir. ▪ Inspect intake line for restriction, kinking, or closed suction valve. ▪ Check clogged strainers or suction filters.
Is it a variable displacement pump?	▪ Check the setting of the pump controller.
Pump-motor coupling is sheared.	▪ Check pressure spikes. ▪ Check coupling misalignment.
Pump is not primed.	▪ Prime the pump after reviewing the relevant instructions.
Pumping elements are severely worn, damaged or seized?	▪ Stuck internal components from varnish in the oil or from rust and corrosion. Varnish indicates the system is running too hot. Rust or corrosion may indicate water in the oil.
Pump is wrongly assembled. **(See Note 1).**	▪ Assemble the pump according to the manufacturer assembly instructions.

Table 4.2- Troubleshooting Chart (T-Pump-01-No Flow out of the Pump)

Note 1: An example of an incorrectly assembled pump is, shown in Fig. 4.1, a vane pump with angled vanes. As shown in figure, Attention should be paid during assembling the edged vanes. Due to the incorrect placement of the vane, pressure force during the pressure stroke will act normally on the edged surface as per Pascal's Law.

Pressure force can be resolved into two components. The vertical component will push the vane down, which will affect the sealing conditions between the vane and the stator. The horizontal component will tilt the vane, which will cause severe friction at the local points shown in the figure.

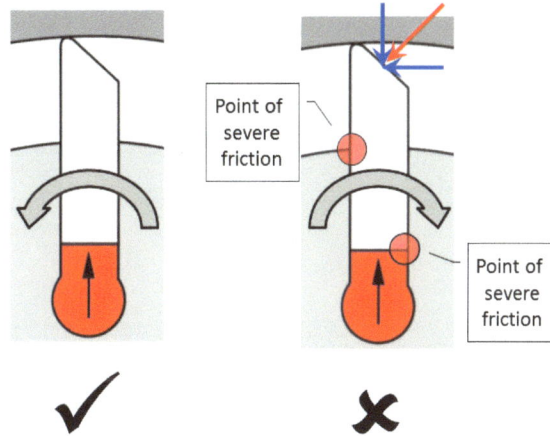

Fig. 4.1 – Correct Direction of Rotation

4.2.2- Low Flow Out of the Pump

Table 4.3 shows the troubleshooting actions when low *flow* out of the pump, before the relief valve, is observed.

T-Pump-02-Low Flow out of the Pump	
Pump is recently changed.	▪ Review the pump ordering code.
Pump rotates at low speed.	▪ Check the drive motor speed.
Pump is driven by 3-phase electric motor.	▪ Check fuses on all three phases of a 3-phase electric motor. If the fuse on one phase is blown the motor may run but will overheat and will not produce full power.
Is it a variable displacement pump?	▪ Check the setting of the pump controller. ▪ Check over pressure for pressure compensated pumps.
Pump cavitation.	▪ Consult Chart: "**T-System-02-Pump Cavitation**".
Pump worn.	▪ Conduct pump performance test **(see Note 1)** to check internal leakage rate and then rebuild or replace the pump accordingly.

Table 4.3- Troubleshooting Chart (T-Pump-02-Low Flow out of the Pump)

Note 1: Figure 4.2 shows the typical pump test results that shows the pump flow **Q** versus working pressure **p**. The figure shows that pump A and pump B both show the ability to achieve certain pressure. Pump B discharges much less flow which means it has more leakage. However, such a test is used to compare the actual pump flow versus the rated flow at certain working pressure. The result of this test indicates the status of the pump if it needs just rebuild or replacement.

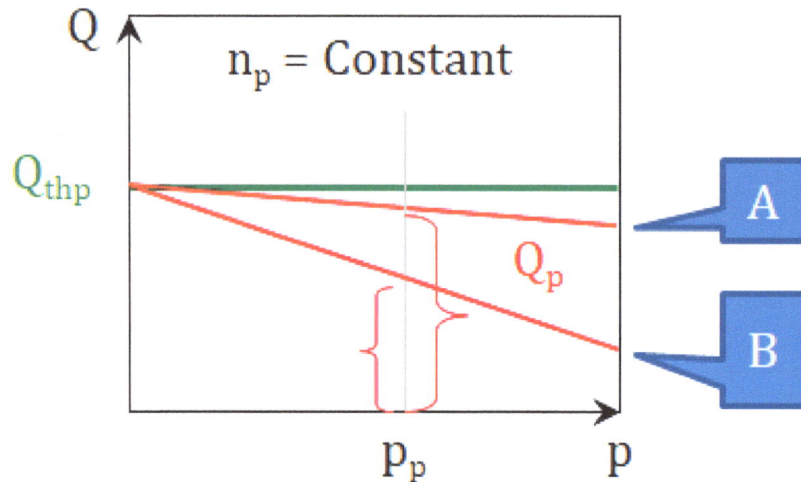

Fig. 4.2 – Pump Test Results

4.2.3- Erratic Flow Out of the Pump

Table 4.4 shows the troubleshooting actions when erratic *flow* out of the pump is observed.

T-Pump-03-Erratic Flow out of the Pump	
Fluid Aeration.	▪ Consult Chart: ▪ **"T-System-01-Fluid Aeration"**.
Pump Cavitation.	▪ Consult Chart: ▪ **"T-System-02-Pump Cavitation"**.
Is it a variable displacement pump?	▪ Check the setting of the pump controller.
Worn or inefficient pump.	▪ Test the pump.
Excessive pump wear.	▪ Consult Chart: ▪ **"T-Pump-10-Excessive Pump Wear"**.

Table 4.4- Troubleshooting Chart (T-Pump-03-Erratic Flow out of the Pump)

4.2.4- Excessive Flow Out of the Pump

Table 4.5 shows the troubleshooting actions when excessive *flow* out of the pump is observed.

T-Pump-04-Excessive Flow out of the Pump	
Pump is oversized.	▪ Review the pump ordering code. ▪ Check if cam ring of a vane pump is worn.
Pump rotates at high speed	▪ Check the drive motor speed.
Is it a variable displacement pump?	▪ Check the setting of the pump controller.

Table 4.5- Troubleshooting Chart (T-Pump-04-Excessive Flow out of the Pump)

4.2.5- No Pressure at the Pump Outlet

Table 4.6 shows the troubleshooting actions when no *pressure* is observed at the pump outlet.

T-Pump-05-No Pressure at the Pump Outlet	
No flow out of the pump?	▪ Consult Chart: ▪ **"T-Pump-01-No Flow out of the Pump"**.
Pressure gauge is faulty?	▪ Replace the pressure gauge.
Pressure relief valve is left fully opened.	▪ Reset the valve.
Is it a variable displacement pump?	▪ Check the setting of the pump controller.
Circuit design issues	▪ Check if the pump is vented through an open or tandem center valve? ▪ Check pump unloading method.

Table 4.6- Troubleshooting Chart (T-Pump-05-No Pressure at the Pump Outlet)

4.2.6- Low Pressure at the Pump Outlet

Table 4.7 shows the troubleshooting actions when low *pressure* is observed at pump outlet.

T-Pump-06-Low Pressure at the Pump Outlet	
Pressure relief valve is set too low.	▪ Reset the valve.
Is it a variable displacement pump?	▪ Check the setting of the pump controller.
External load is below normal or no load?	▪ Check load conditions.
Leaking Pump?	▪ Consult Chart: ▪ **"T-Pump-09-Leaking Pump"**.
Pressure relief valve is at fault?	▪ Consult Chart: ▪ **"T-Valve-03-PCV Troubleshooting"**.

Table 4.7- Troubleshooting Chart (T-Pump-06-Low Pressure at the Pump Outlet)

4.2.7- Erratic Pressure at the Pump Outlet

Table 4.8 shows the troubleshooting actions when low *pressure* is observed at pump outlet.

T-Pump-07-Erratic Pressure at the Pump Outlet	
Erratic flow out of the pump?	▪ Consult Chart: ▪ **"T-Pump-03-Erratic Flow out of the Pump"**.
Excessive pump wear?	▪ Consult Chart: ▪ **"T-Pump-10-Excessive Pump Wear"**.
Pressure relief valve is at fault (worn or sticking relief valve)?	▪ Consult Chart: ▪ **"T-Valve-03-PCV Troubleshooting"**.
A pump works with an accumulator in parallel?	▪ Check the accumulator for gas leak or damage of accumulator piston, bladders, or diaphragm.
Backup pressure relief valve set near setting of a variable pump controller.	▪ Set relief valve at 10% higher than the variable pump compensator **(See Note 1)**.
Shuddering of overrunning load controlled by PO Check?	▪ Check sizing of the PO Check ▪ Valve and review circuit design **(See Note 2)**.

Table 4.8- Troubleshooting Chart (T-Pump-07-Erratic Pressure at the Pump Outlet)

Note 1: As shown in Fig. 4.3, a backup relief valve may be used as a redundancy to work in case if the ump controller failed. This backup relief valve should be set at least 10 bar (150 psi) higher than the pump compensator. In case if the backup relief valve is set near the compensator, pump controller may frequently actuate resulting in erratic flow and pressure.

Note 2: As shown in Fig. 4.4, one way to control overrunning load is to use pilot operated check valve. In case if the size of the pump, the cylinder, and the pilot operated check valve aren't compromisingly sized, load shuddering may occur during lowering the load

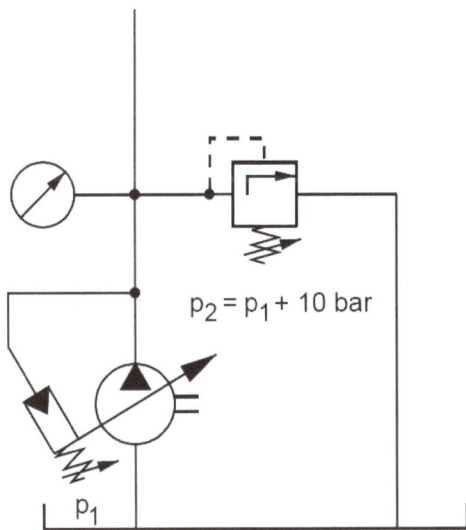

Fig. 4.3- Pressure Compensated Pump with a Backup Pressure Relief Valve

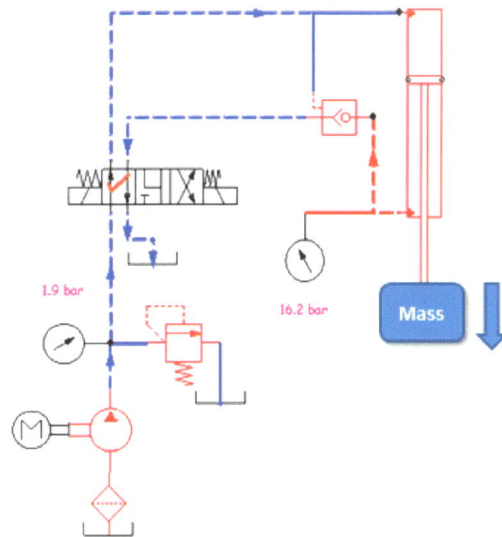

Fig. 4.4- Pilot Operated Check Valve to Control Overrunning Load

4.2.8- Excessive Pressure at the Pump Outlet

Table 4.9 shows the troubleshooting actions when excessive *pressure* is observed.

T-Pump-08-Excessive Pressure at the Pump Outlet	
Pressure relief valve is set too high.	▪ Reset the valve.
Is it a variable displacement pump?	▪ Check the setting of the pump controller.
External load is above normal?	▪ Check load conditions.
Pressure relief valve is at fault.	▪ Consult Chart: ▪ "T-Valve-03-PCV Troubleshooting".

Table 4.9- Troubleshooting Chart (T-Pump-08-Excessive Pressure at the Pump Outlet)

4.2.9- Leaking Pump

Table 4.10 shows the troubleshooting actions when *leaking* pump is observed.

T-Pump-09-Leaking Pump	
Pump is very hot?	<ul><li>Consult Charts:</li><li>**"T-Unit-03-Excessively Hot Unit".**</li><li>**"T-System-04-Excessive System Heat".**</li></ul>
Pump is over pressurized?	<ul><li>Check working pressure versus max allowable pressure for the pump.</li><li>Consult Chart: **"T-Pump-08: Excessive Pressure at the Pump Outlet".**</li></ul>
Pump is noticeably noisy and vibrating?	<ul><li>Consult Charts:</li><li>**"T-Unit-02-Noisy Unit".**</li><li>**"T-Pump-12-Excessive Pump Noise and Vibration".**</li><li>**"T-System-03-Excessive System Noise and Vibration".**</li></ul>
Lose or broken pressure lines? Damaged thread of pump ports? Use of incorrect fittings?	<ul><li>Tighten or replace the line.</li><li>Repair/replace pump housing.</li><li>Use standard fittings.</li></ul>
Case drain restricted or too small?	<ul><li>Check case drain pressure.</li><li>Size the drain line properly.</li><li>Eliminate kinks and bends.</li></ul>
Pump housing cracked due to mechanical stress or vibration?	<ul><li>Replace the pump.</li><li>Tighten pump housing properly.</li><li>Isolate pump housing from vibration.</li></ul>
Shaft seal leak? Contamination between shaft and seal?	<ul><li>Consult Chart:</li><li>**"T-Seals-01-Seal Troubleshooting".**</li></ul>
Excessive pump wear?	<ul><li>Consult Charts:</li><li>**"T-Pump-10-Excessive Pump Wear".**</li></ul>

Table 4.10- Troubleshooting Chart (T-Pump-09-Leaking Pump)

4.2.10- Excessive Pump Wear or Inside Parts Broken

Table 4.11 shows the troubleshooting actions when excessive pump *wear* is observed.

T-Pump-10-Excessive Pump Wear	
Fluid Aeration?	<ul><li>Consult Chart:</li><li>**"T-System-01-Fluid Aeration"**.</li></ul>
Pump Cavitation?	<ul><li>Consult Chart:</li><li>**"T-System-02-Pump Cavitation"**.</li></ul>
Abrasive dirt in the fluid?	<ul><li>Drain and flush the system thoroughly.</li><li>Replace filter element.</li><li>Investigate sources of wear base on fluid analysis.</li></ul>
Fluid viscosity too low or too high?	<ul><li>Check the working temperature versus recommended fluid viscosity.</li><li>Act accordingly (adjust working temperature or change fluid).</li></ul>
Higher water content in fluid?	<ul><li>Investigate sources of water penetration to the system and then act accordingly.</li></ul>
Working pressure is above normal?	<ul><li>Check working pressure versus max allowable pressure for the pump.</li><li>Consult Chart: **"T-Pump-08: Excessive Pressure at the Pump Outlet"**.</li></ul>
Pump-coupling misalignment or drive belt is too tight?	<ul><li>Check pump shaft alignment with the coupling.</li><li>Check belt tension (if found).</li><li>Check maximum radial and axial load on shaft.</li></ul>

Table 4.11- Troubleshooting Chart (T-Pump-10-Excessive Pump Wear or Inside Parts Broken)

4.2.11- Air Leaks into Pump

Table 4.12 shows the troubleshooting actions when *air leak* into pump is observed.

T-Pump-11-Air Leaks into Pump	
Is the reservoir fluid level too low?	▪ Follow the guidelines to make up the oil in the reservoir to the specified level.
Leaking fitting in the intake line?	▪ Tighten the leaking fitting on intake line.
Pump shaft seal worn or damaged?	▪ Replace the pump shaft seal.

Table 4.12- Troubleshooting Chart (T-Pump-11-Air Leaks into Pump)

4.2.12- Excessive Pump Noise and Vibration

Table 4.13 shows the troubleshooting actions when *noise and vibration* are observed.

T-Pump-12-Excessive Pump Noise and Vibration	
Pump Cavitation (is vacuum in the pump intake below recommendations)?	▪ Consult Chart: ▪ **"T-System-02-Pump Cavitation".**
Excessive pump wear or inside parts broken?	▪ Consult Chart: ▪ **"T-System-10-Excessive Pump Wear".**
Wrong direction of pump rotation?	▪ Check electrical wiring to electric motor.
Working pressure is too high?	▪ Check working pressure vs. maximum allowable pressure of the pump.
Pump housing bolts not tightened properly.	▪ Check and tighten according to manufacturer recommendations.
Noise damping cushions worn.	Check and replace if needed.
Coupling or other transmission elements are wrongly aligned or lose?	Check and resolve accordingly.
Is the noise due to pressure ripples from a positive displacement pump?	▪ Consider installing a small accumulator downstream the pump OR Shock suppressor.

Table 4.13- Troubleshooting Chart (T-Pump-12-Excessive Pump Noise and Vibration)

4.3- Hydraulic Pumps Failure Analysis

Hydraulic pumps are generally the most expensive components in hydraulic systems. They have the highest reliability risk, the highest contamination sensitivity, and the ability to cause chain-reaction failures. In other words, when a pump starts to fail, it generates debris into the fluid flow downstream of the pump. If there is no good filtration, this debris moves on to other components like valves, actuators, and can lead to damage in those components as well.

4.3.1- Pump Failures due to Contamination

To get a better sense of the effect of *contamination*, Table 4.14 shows pump failure frequency due to different reasons. It is evident that majority of pump failures are due to fluid contamination. Figure 4.5 provides a practical effect of cleanliness level on pump lifetime.

Source	Failure Frequency (%)
Contamination	80
Installation	12
Manufacturing Defects	6
Design	2

Table 4.14- Pump Failure Frequency due to Various Reasons

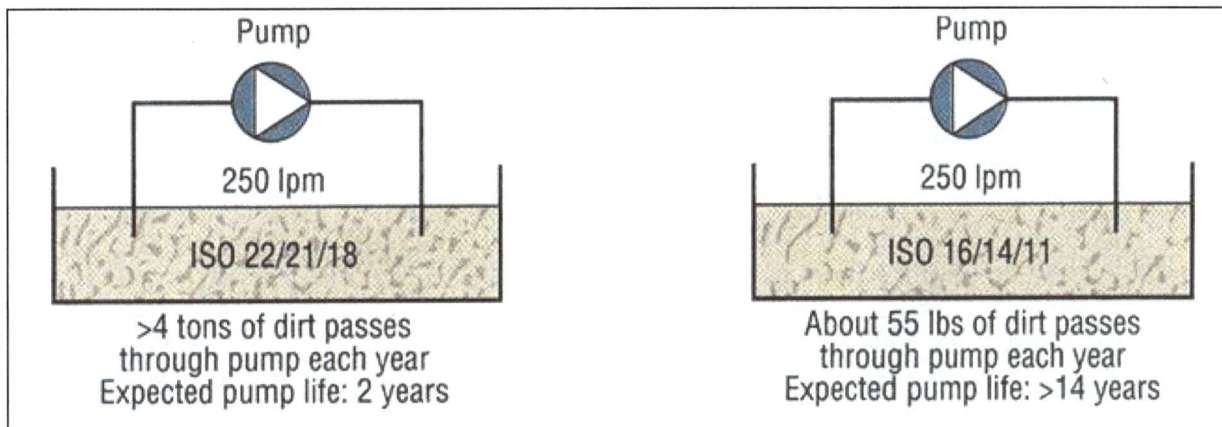

Fig. 4.5- Effect of Cleanliness Level on a Pump Lifetime (Hydraulic & Pneumatic Magazine)

Figure 4.6 shows opposing moving surfaces within hydraulic pumps that are commonly affected by abrasive wear as follows:

- **In Gear Pumps and Motors:**
  - The radial clearance between opposite teeth of a gear pump or motor and between the tip of the teeth and the housing.
  - The side clearance between the face of the gears and the bearing plates.

- **In Vane Pumps and Motors:**
  - The radial clearance between the tip of each vane and the cam ring.
  - The side clearance between the body of the vane and the rotor.

- **In Piston Pumps and Motors:**
  - The clearance between the cylinder block and the valve plate.
  - The clearance between each piston and its piston chamber.
  - The clearance between the spherical head of each piston and its slipper pad.
  - The clearance between the slipper pads and the swash plate.

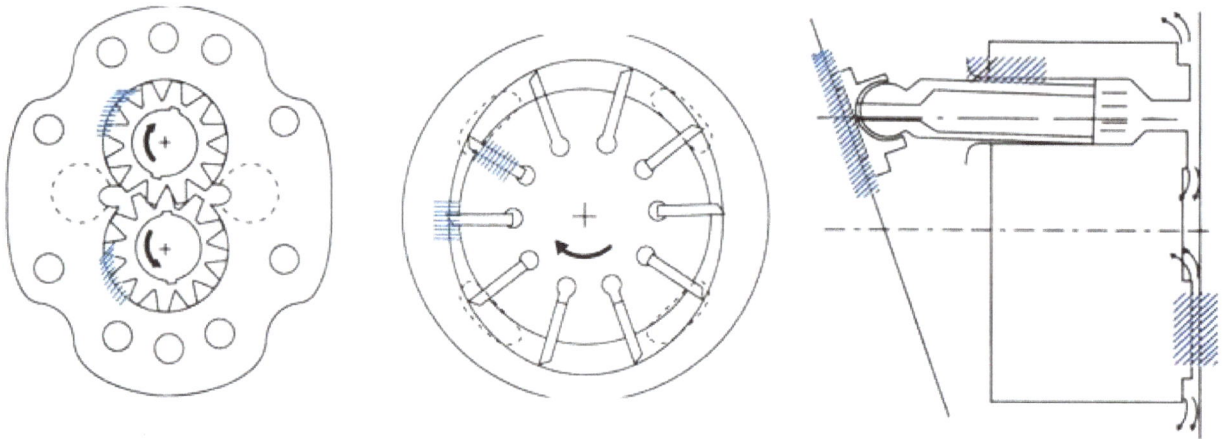

Fig. 4.6- Commonly Worn Areas within Hydraulic Pumps and Motors (Courtesy of Pall)

4.3.1.1- Failures due to Contamination in Gear Pumps

Housing Failure due to Contamination: As shown in Fig. 4.7, a positive displacement pump is an unbalanced pump because it has high pressure at the discharge side and low pressure at the suction side. Therefore, bearing load is not evenly distributed along the bearing circumference, and rather concentrated on one side. Severe *bearing* wear occurs in presence of particulate contamination. Figure 4.8 shows an example of gear pump housing failure due to contamination.

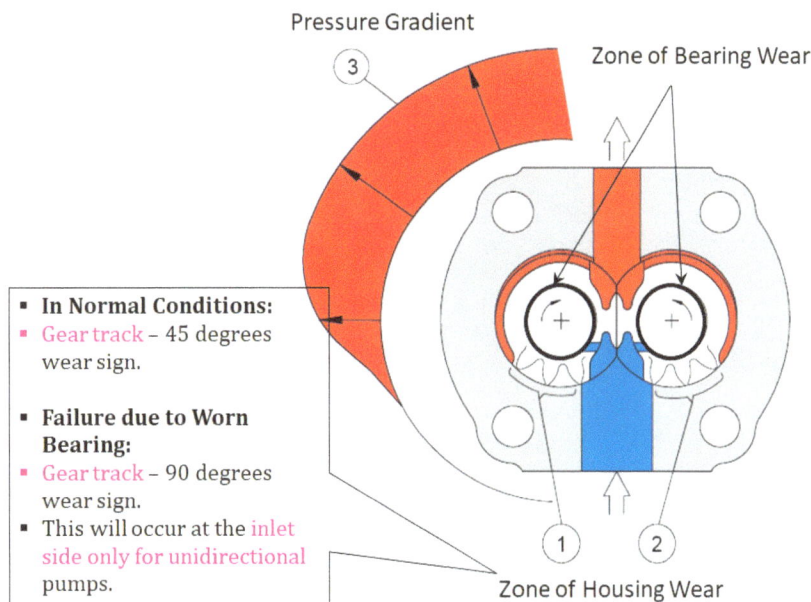

Pressure Gradient

Zone of Bearing Wear

- **In Normal Conditions:**
- Gear track – 45 degrees wear sign.

- **Failure due to Worn Bearing:**
- Gear track – 90 degrees wear sign.
- This will occur at the inlet side only for unidirectional pumps.

Zone of Housing Wear

Fig. 4.7- Wear Zones in Gear Pump and Motor Bearings

- **In Normal Conditions:**
- Gear track - provides low gear tip clearance and high volumetric efficiency.
- Nominal depth = .008" = (0.203 mm)
- Should not exceed .015" = (0.381 mm)

- **Failure due to Contamination:**
- Gear track – 90 degrees wear sign.
- Contamination by fine particles will cause the gear track to be gray with a sandblasted appearance
- This will occur at the inlet side only for unidirectional pumps.

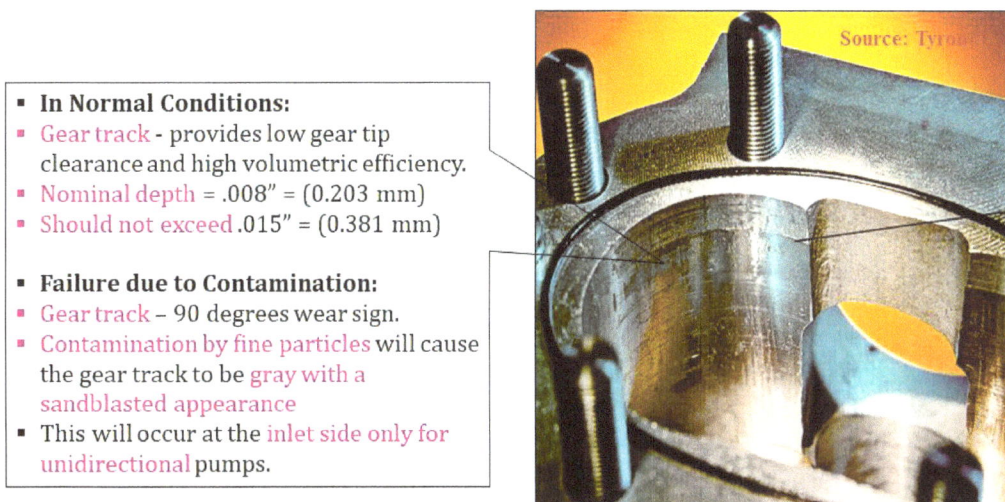

Source: Tyrone

Fig. 4.8- Example of Gear Pump Housing Failure due to Contamination

Pressure Plate Failure due to Contamination: As shown in Fig. 4.9, an external gear pump contains two *pressure plates*, one on each side of the gears to control the side clearance with the increase of working pressure so that internal leakage is accordingly controlled. These plates are also used as bearing for the gears. Figure 4.10 shows possible failure due to contaminations that can find its way between the *pressure plates* and the gears. Usually, pressure plates are made of material lighter than the material of the gears, so they wear on behalf of the gear. That is why they also referred to as wearing plates.

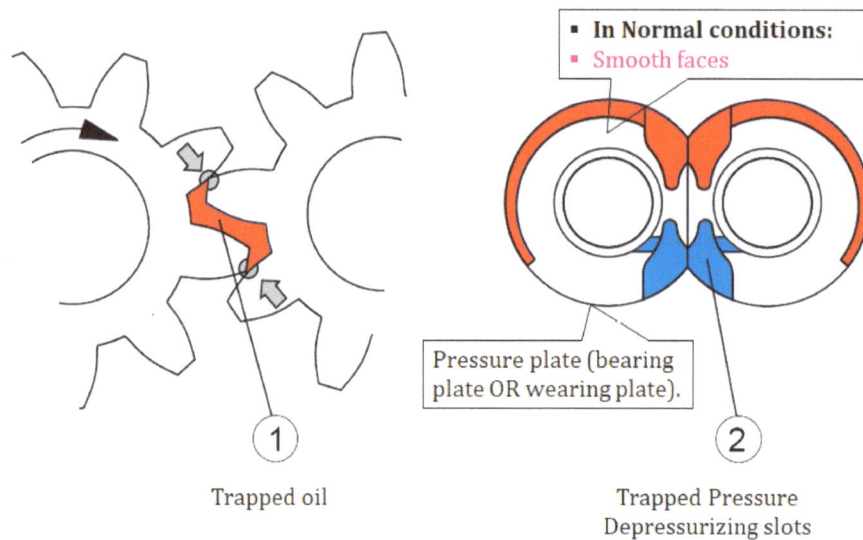

- **In Normal conditions:**
- Smooth faces

Pressure plate (bearing plate OR wearing plate).

1 Trapped oil

2 Trapped Pressure Depressurizing slots

Fig. 4.9- Gear Pump Pressure Plates

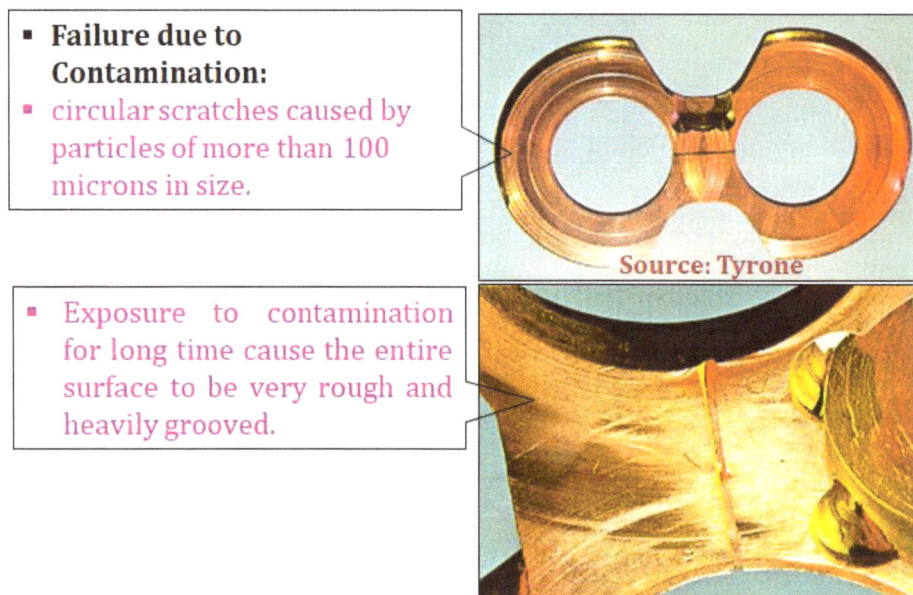

- **Failure due to Contamination:**
- circular scratches caused by particles of more than 100 microns in size.

Source: Tyrone

- Exposure to contamination for long time cause the entire surface to be very rough and heavily grooved.

Fig. 4.10- Example of Gear Pump Pressure Plate Failure due to Contamination

Input Shaft Failure due to Contamination: As shown in Fig. 4.11, gear pump input *shaft* looks shiny and smooth when you pass your nail over it. The figure shows that abrasive contaminates cause circular scratches between the shaft and bearings or between the shaft and the seal lip.

- ▪ **In Normal conditions:**
- ▪ Smooth shaft.

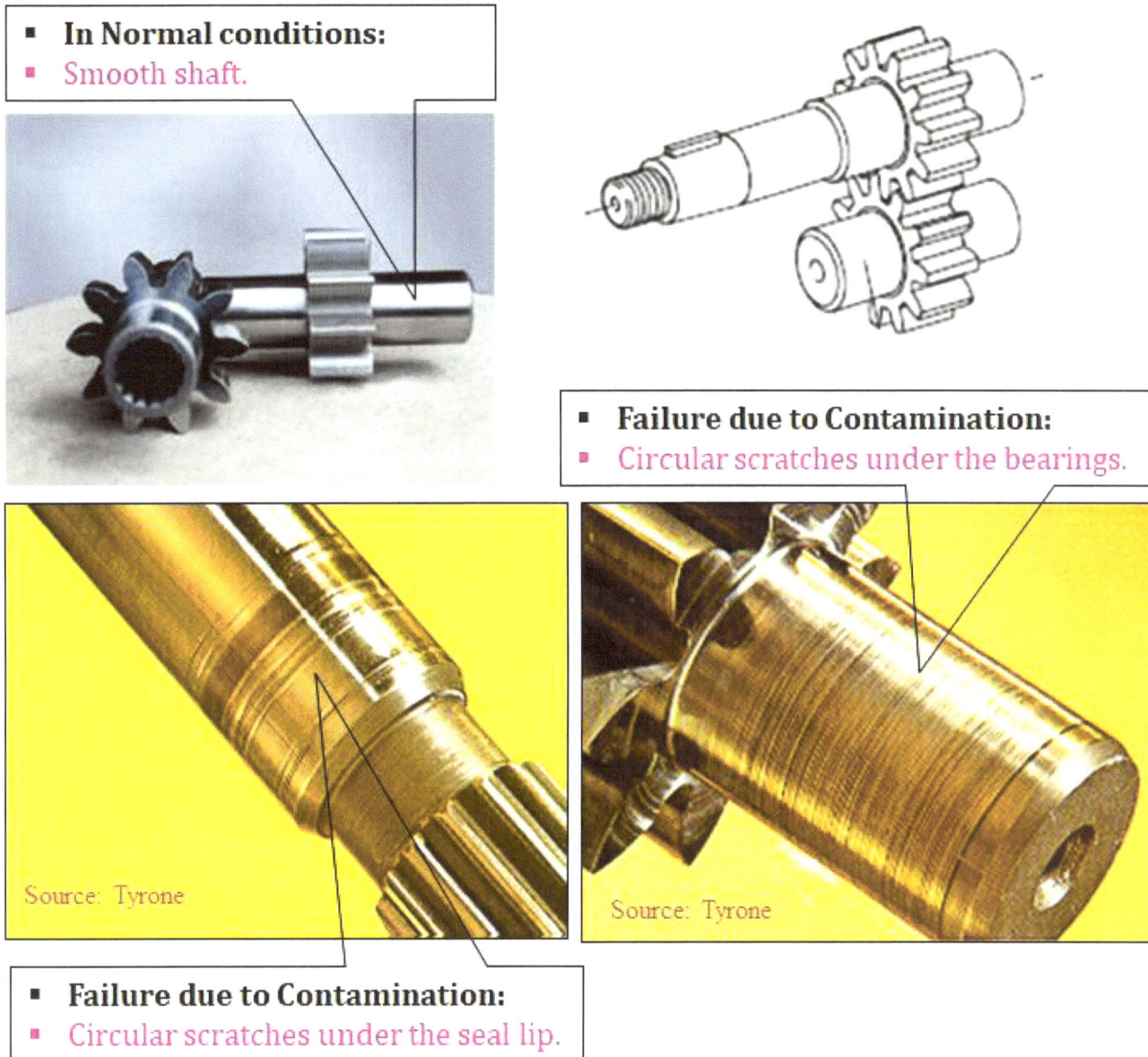

- ▪ **Failure due to Contamination:**
- ▪ Circular scratches under the bearings.

Source: Tyrone

Source: Tyrone

- ▪ **Failure due to Contamination:**
- ▪ Circular scratches under the seal lip.

Fig. 4.11- Example of Gear Pump Input Shaft Failure due to Contamination

Bearing Failure due to Contamination: Figure 4.12 shows examples of bearing failure due to particulate contamination. The figure shows a destroyed raceway (1) of a ball bearing caused by particulate contamination, a chip (2) embedded in a surface of an anti-friction *bearing* and a destroyed roller bearing (3) in a piston pump.

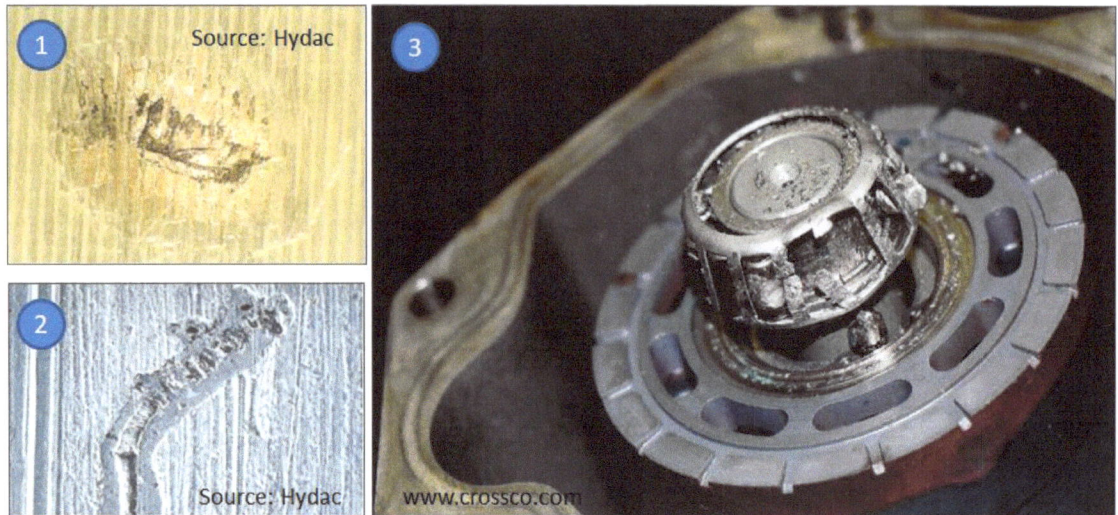

Fig. 4.12- Examples of Bearing Failures due to Particulate Contamination

4.3.1.2- Failures due to Contamination in Vane Pumps

Wear Plate Failure due to Contamination:
Figure 4.13 shows the damage caused by large metal objects jammed between the rotor and the *wear plate*. This wear plate is unrepairable and must be replaced.

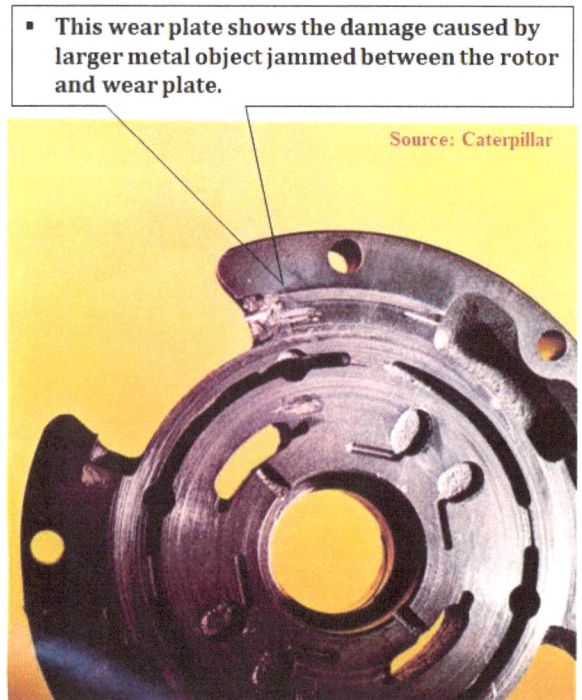

- This wear plate shows the damage caused by larger metal object jammed between the rotor and wear plate.

Source: Caterpillar

Fig. 4.13- Examples of Wear Plate Failures due to Particulate Contamination

Vanes Failure due to Contamination: Figure 4.14 shows various failures in *vane* pumps due to fluid contamination. As shown in the figure, failures could range from slight condition to heavy wear and even jamming the vanes inside the rotor.

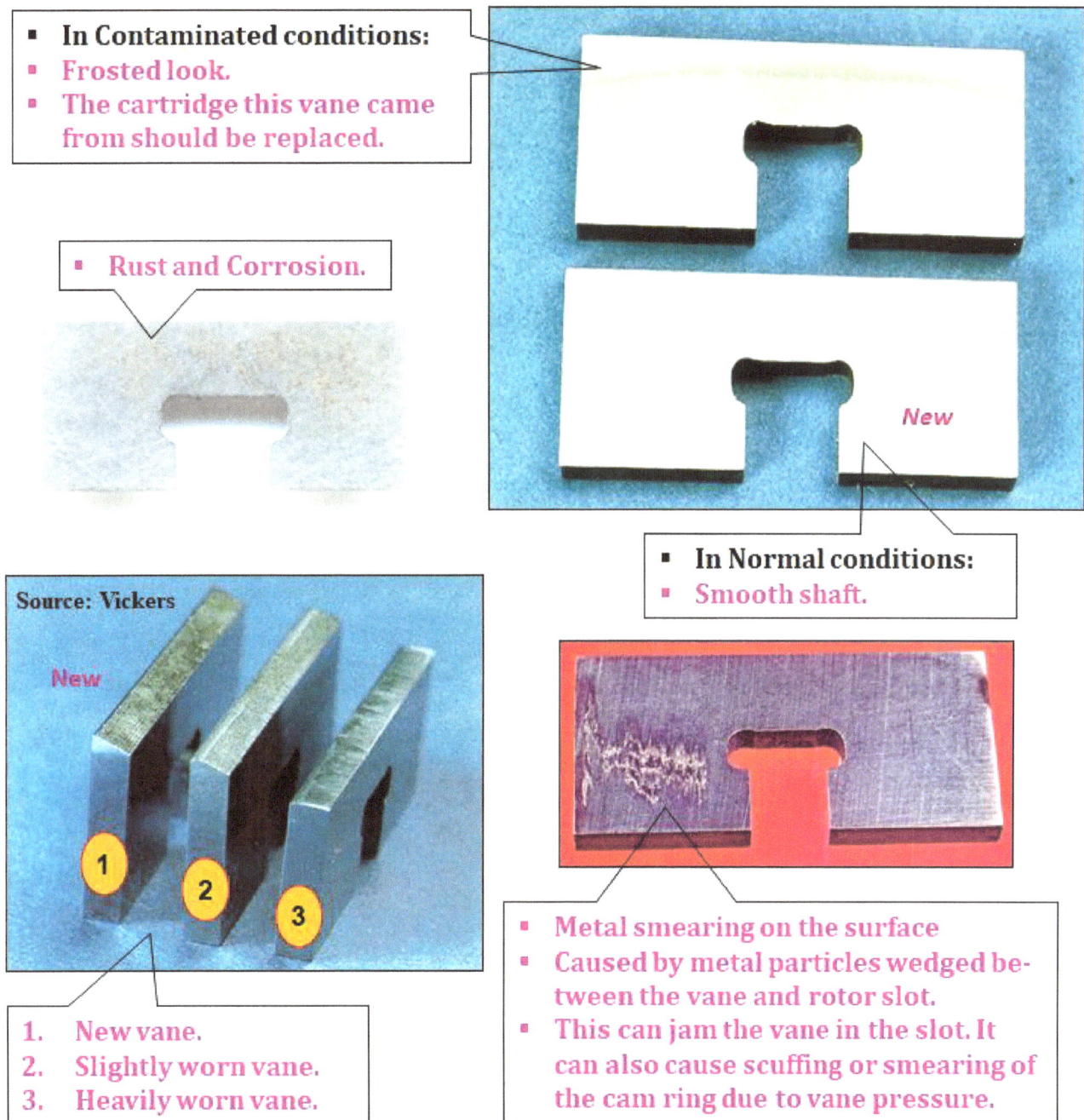

- **In Contaminated conditions:**
- **Frosted look.**
- **The cartridge this vane came from should be replaced.**

- **Rust and Corrosion.**

New

- **In Normal conditions:**
- **Smooth shaft.**

Source: Vickers

New

1. **New vane.**
2. **Slightly worn vane.**
3. **Heavily worn vane.**

- **Metal smearing on the surface**
- **Caused by metal particles wedged between the vane and rotor slot.**
- **This can jam the vane in the slot. It can also cause scuffing or smearing of the cam ring due to vane pressure.**

Fig. 4.14- Examples of Vanes Failures due to Particulate Contamination

Cam Ring Failure due to Contamination: Figure 4.15 shows a worn cam ring. It is to be noted that, unlike other pumps, worn *cam ring* in a vane pump could result in increasing the pump flow because the cam ring becomes larger. However, worn cam ring must be replaced.

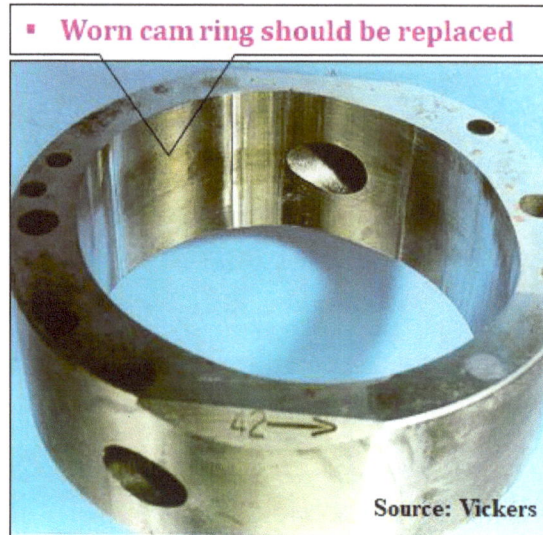

Fig. 4.15- Examples of Cam Ring Failures due to Particulate Contamination

Rotor Failure due to Contamination: Figure 4.16 shows a scored *rotor* due to contaminants jammed between the cam ring and the rotor.

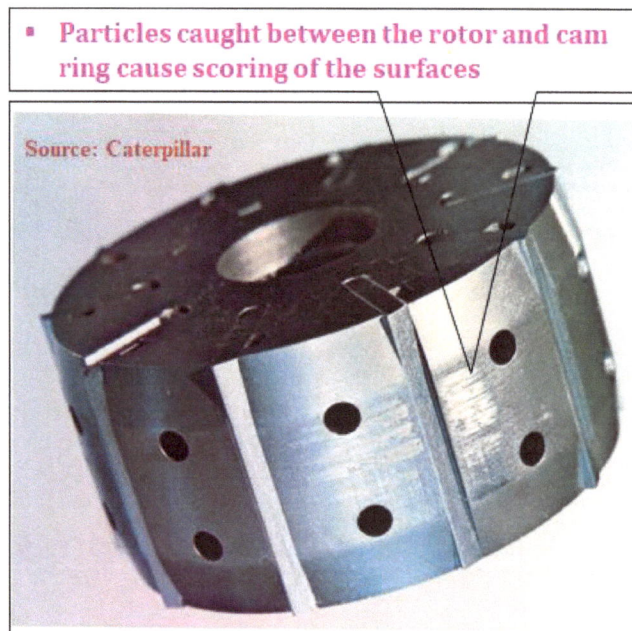

Fig. 4.16- Examples of Rotor Failures due to Particulate Contamination

4.3.1.3- Failures due to Contamination in Piston Pumps

Blocking Lubrication Passages: Figure 4.17 shows how particulate contamination causes blocking the *lubrication hole* in an axial piston swash plate pump.

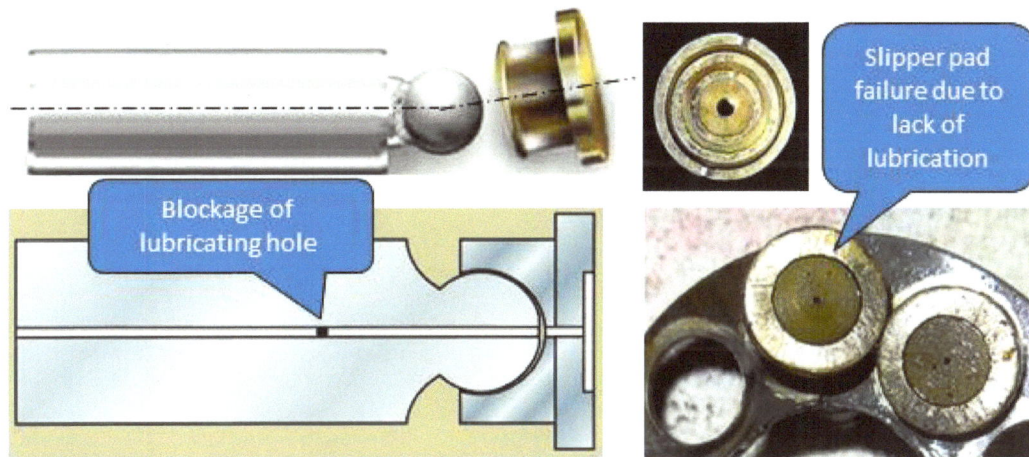

Fig. 4.17- Commonly Worn Areas within Hydraulic Pumps and Motors (Courtesy of Pall)

Cylinder Block Failure due to Contamination: Figure 4.18 shows *cylinder block* top surface scoring as an evidence of contamination. In such cases the cylinder block can be re-lapped or reground .005" to .015" (0.127 – 0.381 mm).

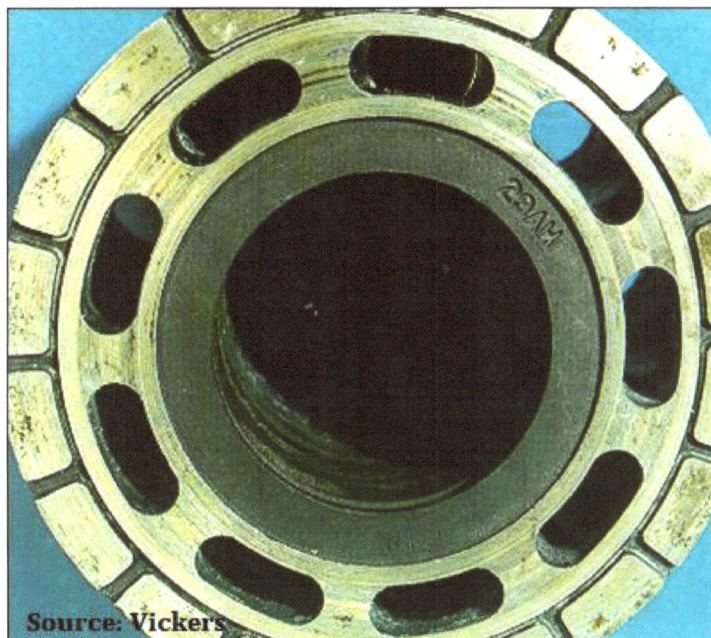

Fig. 4.18- Example of Cylinder Block Failure due to Contamination

Failure of Valve Plate due to Contamination: Figure 4.19 shows *valve plate* surface can become scored due to a number of factors, including contamination. Scored surfaces like this one can be resurfaced up to .015" (0.381 mm) with making sure *Silencing Grooves* are not affected. Figure 4.20 shows when valve plate was badly damaged by a large piece of foreign material. It can't be resurfaced or reused.

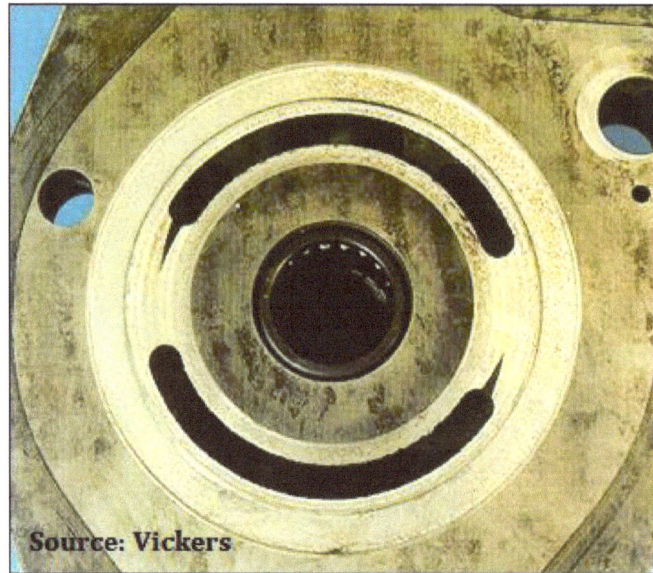

Fig. 4.19- Example of Valve Plate Failure due to Contamination

Fig. 4.20- Example of Heavily Damaged Valve Plate due to Contamination

Failure of Piston due to Contamination: Figure 4.21 shows damaged *piston head* due to contamination. Piston head can become loose from the shoe as a result of severe scoring and pitting from contamination. Old pistons should never be re-used. Figure 4.22 shows the piston on the left (#1) is in relatively good condition and can be used again. the piston on the right (#2) has obvious damage and should be replaced. It can't be reused.

Fig. 4.21- Example of Piston Head Failure due to Contamination

Fig. 4.22- Example of Piston Body Seizure due to Contamination

Exceeding Cleanliness Level: Figure 4.23 shows examples of piston pump failures due to particulate contamination. Damage occurs when particulate *contamination* level (ISO 4406) exceeds manufacturers recommendations. The figure shows worn/broken slipper pads, retaining plates, and valve plate.

Fig. 4.23- Examples of Piston Pumps Failure due to Particulate Contamination

4.3.2- Pump Failures due to Overheating

Everyone knows that contamination can be catastrophic to a hydraulic system. But *overheating* can also be detrimental to hydraulic fluid and the components within that system.

4.3.2.1- Failures due to Overheating in Gear Pumps

Failure of Pressure Plate due to Overheating: Figure 4.24 shows a damaged *pressure plates* when the temperature is raised above 300 ⁰F. The figure shows the entire pressure plate is coated with a black layer. Despite the surface shows very little damage, but it can't be reused.

Fig. 4.24- Example of Pressure Plate Failure due to Overheating

Failure of Input Shaft due to Overheating: Figure 4.25 shows *shaft*, gears, and bearing surfaces will be black all over. Shaft will show some bright circles, but no real grooves. However, these parts shall not be used. Exposure to excessive heat will also discolor the ends of the gear teeth near the gear face. With continued operation the gear face and thrust plate will start to weld together. Friction generated from continued operation could eventually result in a seized pump.

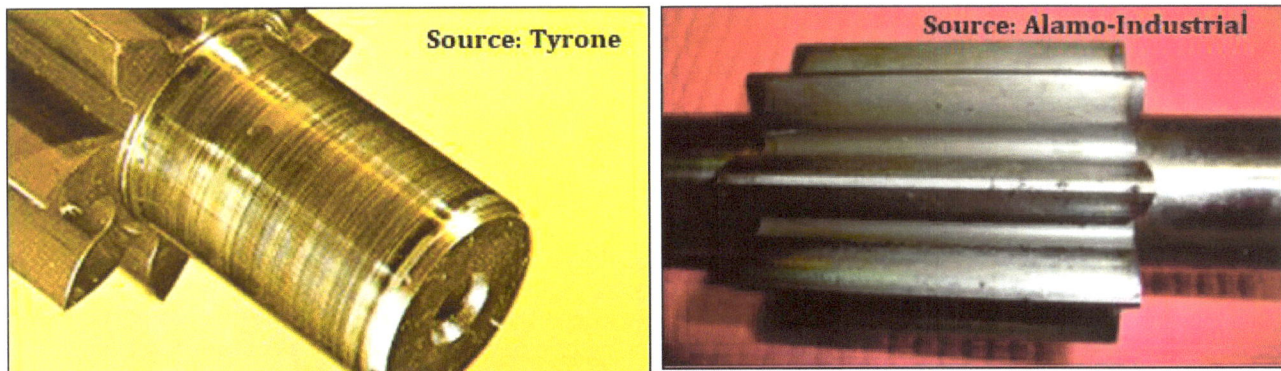

Fig. 4.25- Example of Input Shaft and Gears Failure due to Overheating

Failure of Seal Strip due to Overheating: Figure 4.26 shows a broken *seal strip* that is hardened and became brittle due to overheating.

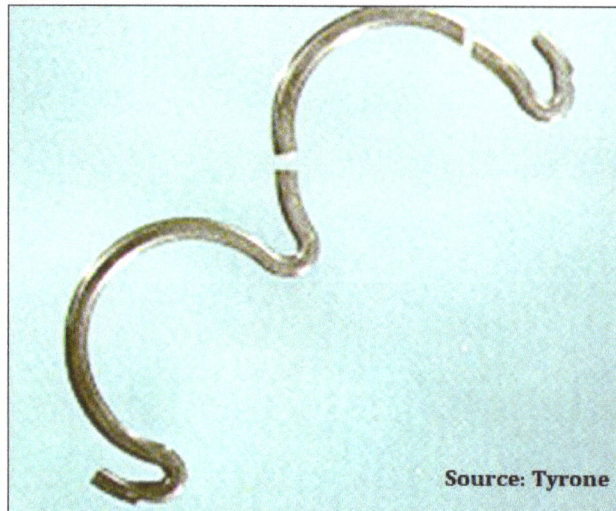

Fig. 4.26- Example of Seal Strip Failure due to Overheating

4.3.2.2- Failures due to Overheating in Vane Pumps

Failure of Cam Rings due to Overheating: Figure 4.27 shows the *cam ring* on the left (1) has mild rippling and can be re-polished. If the local temperature was so high, vane tips literally fused into the came ring. The cam ring on the right (2) has decomposed badly from extreme heat and shows evidence of pump seizure. Cam ring in this condition cannot be reworked.

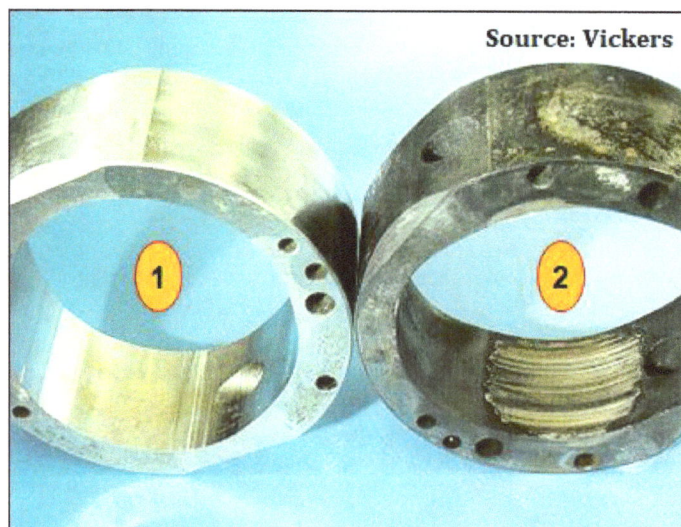

Fig. 4.27- Example of Cam Ring Seizure Damage due to Overheating

Failure of Cam Rings, Rotor, and Bearing due to Overheating: Figure 4.28 shows, on the left side, *discoloring* of a cam ring, a rotor, and a bearing due to overheating. Despite that no signs of cracks or grooves were shown, the entire cartridge should be replaced. On the right side, severe overheating results in lack of lubrication, dry run, and *seizure* damage.

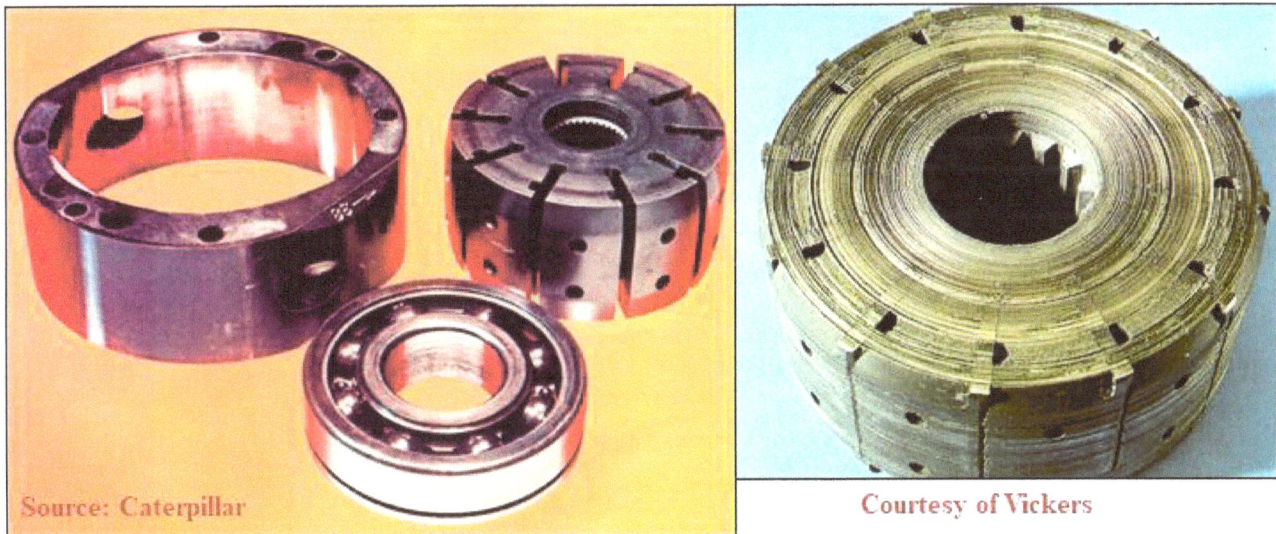

Fig. 4.28- Example of Cam Ring, Rotor, and Bearing due to Overheating

Failure of Valve Plate due to Overheating: Figure 4.29 shows the *valve plate* on the left (1) displays a typical amount of discoloration. The discoloration alone should have no effect on the pump operation. Darkening and erosion on valve plate the right (2) is seen as a result of excessive system temperature. such parts can't be reused.

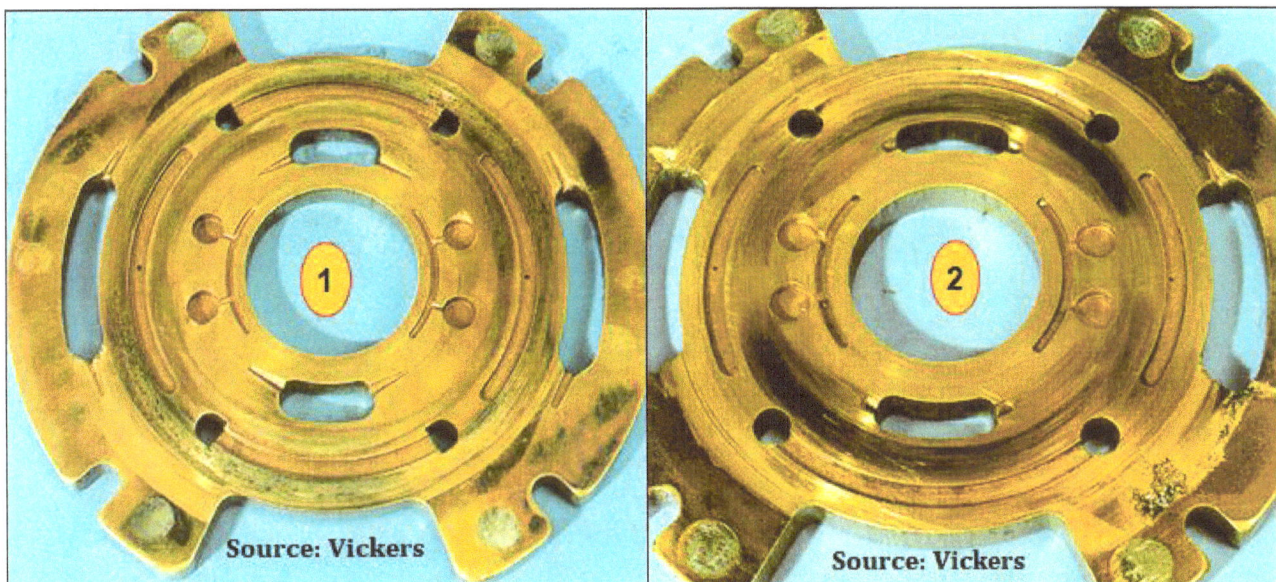

Fig. 4.29- Example of Valve Plate Failure due to Overheating

4.3.2.3- Failures due to Overheating in Piston Pumps

Failure of Cylinder Block due to Overheating: Figure 4.30 shows a *cylinder block* is coated by black layers due to working under excessive temperature for long time. Again, such parts can't be reused because the material properties are changed as it is heat treated.

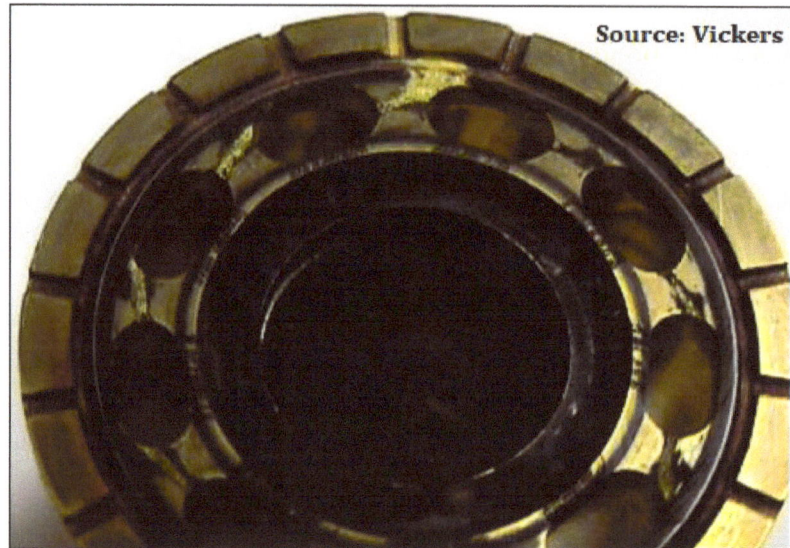

Fig. 4.30- Example of Cylinder Block Failure due to Overheating

Failure of Pistons and Slipper Pads due to Overheating: Excessive system heat can cause hydraulic fluid to break down, resulting in lost viscosity, thinner fluid, and more damage that can cause catastrophic failure to internal components, such as this *piston* and *slipper pad* shown in Fig. 4.31.

Fig. 4.31- Example of Piston and Slipper Pad Failure due to Overheating
(Courtesy of Insane Hydraulics)

4.3.3- Pump Failures due to Cavitation

4.3.3.1- Failures due to Cavitation in Gear Pumps

Failure of Pump Housing due to Cavitation: Figure 4.32 shows various *housing* failures as a result of cavitation development.

Fig. 4.32- Example of Gear Pump Housing Failure due to Cavitation

Failure of Gears due to Cavitation: Figure 4.33 (left) shows physical wear in *gears* near the outside diameter due to *cavitation.* The figure also shows (right) that running the pump at High Speed caused the pump to Cavitate, overheat, rapidly wear, erode, and eventually caused a hardened tooth to break off one of the gears. Being a solid the broken tooth jammed the rest of the gear pump and caused the whole pump to almost instantly stop. Because the pump still engaged with the driving motor, the input shaft of the pump then sheared off.

Fig. 4.33- Example of Gear Failure due to Cavitation

Failure of Pump Pressure Plates due to Cavitation: Figure 4.34 shows physical damage on *pressure plates* due to collapse of bubbles. Cavitation is shown on the thrust plate on the pressure outlet side of the pump in the form of pitting in the thrust plate surface. Discoloration of this thrust plate may also indicate that excessive heat due to lack of oil supply was also present.

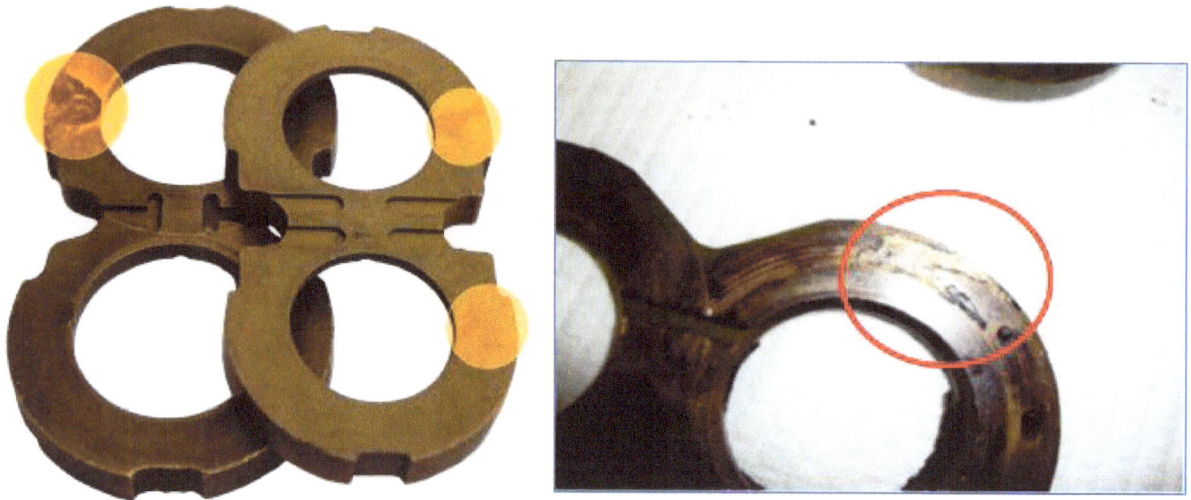

Fig. 4.34- Example of Thrust Plate Failure due to Cavitation (www.alamo-industrial.com)

Failure of Lobe Pump due to Cavitation: Figure 4.35 shows physical damage in housing and rotating elements in *a Lobe* Pump due to cavitation.

Fig. 4.35- Example of Failure in a Lobe Pump due to Cavitation (www.bonvepumps.com)

4.3.3.2- Failures due to Cavitation in Vane Pumps

Failure of Valve Plate due to Cavitation: Sever erosion damage on *valve plate* shown in Fig. 4.36 is caused by collapsed air bubbles. This plate cannot be resurfaced.

Fig. 4.36- Example of Valve Plate Failure due to Cavitation

Failure of Vanes and Cam ring due to Cavitation: Figure 4.37 shows extreme *vane* damage caused by cavitation. The figure shows also a *cam ring* is severely chopped and worn.

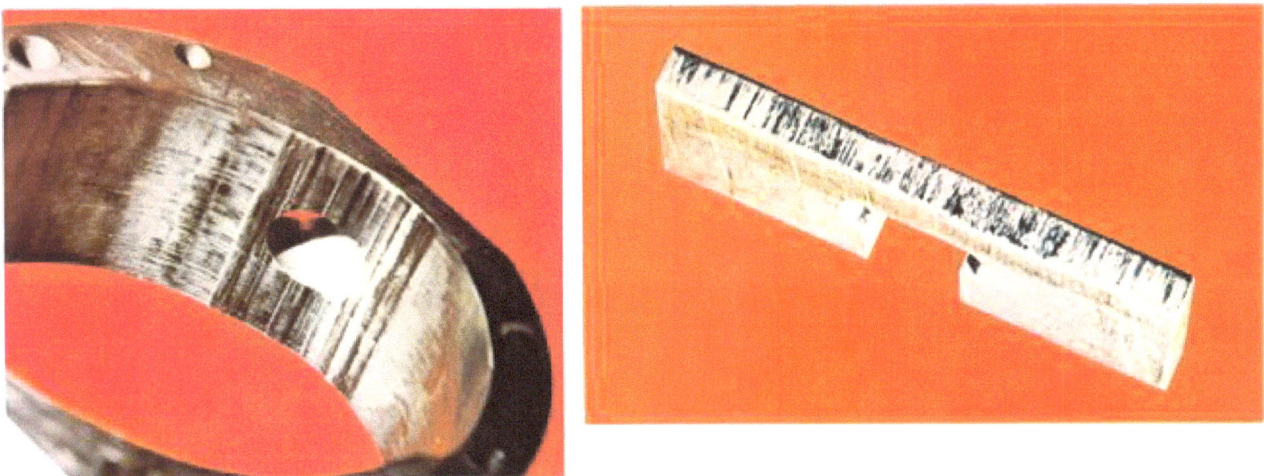

Fig. 4.37- Example of Vane and Cam Ring Failure due to Cavitation

4.3.3.3- Failures due to Cavitation in Piston Pumps

Failure of Cylinder Block due to Cavitation: Figure 4.38 shows the traces of cavitaion damages inside the *piston chambers* of a *cylinder block* in a piston pump.

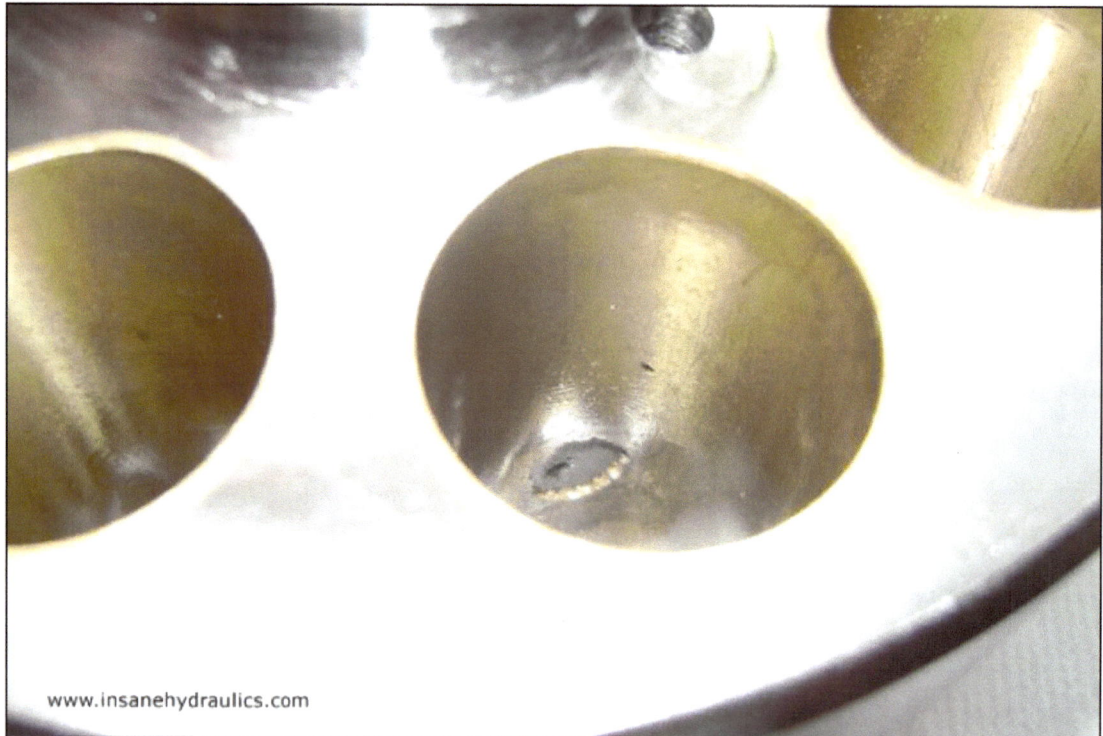

Fig. 4.38- Example of Cavitation Damages in a Cylinder Block of a Piston Pump

Failure of Valve Plate due to Cavitation: Figure 4.39 shows typical cavitation physical damage near the silencing grooves on the valve plate. These *valve plates* are beyond repair.

Fig. 4.39- Example of Valve Plate Failure due to Cavitation

Slipper Pads Lift and Roll due to Cavitation: Some machine owners continue to operate their hydraulics even when excessive noise and vibration is present. As shown in Fig. 4.40, such profound wear can mean only one thing, the shoeless piston heads were hammering the swash plate. During suction stroke, *slipper pads* are lifted and rolled in place when suction force exceeds the spring force that holds the slipper pads against the *swash plate*. Rolling the slipper pads grind the swash plate in circles as shown on the left side. Severe suction forces cause the slipper pads to lift the slipped pads apart from the swash plate. Once the piston enters the discharge stroke, slipper pads strike the swash plate harshly causing the shown damage on the right side. This pate can't be reused and must scrapped.

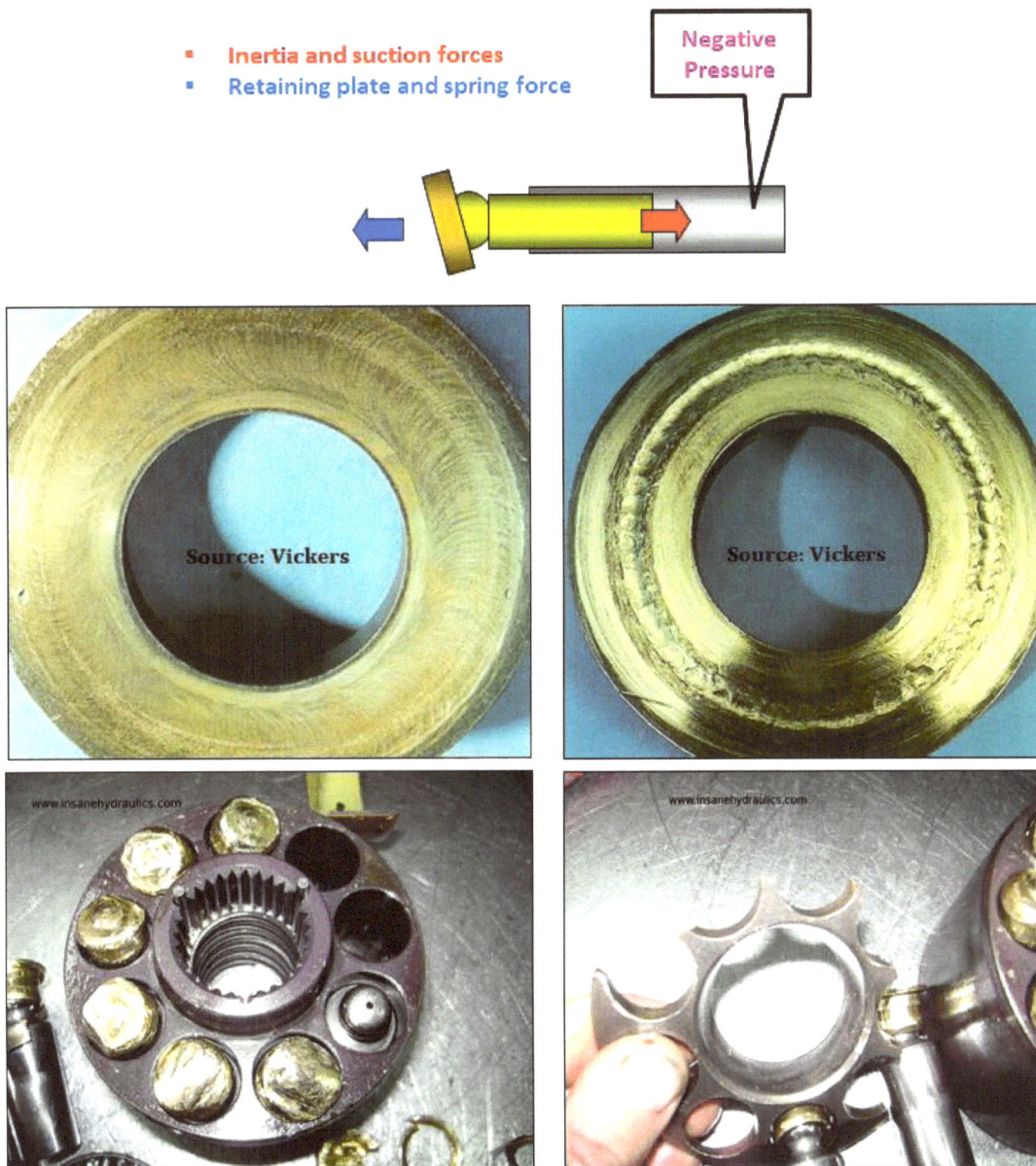

Fig. 4.40- Slipper Pads Lifting and Rolling due to Cavitation

4.3.4- Pump Failures due to Fatigue Stress

Pump Shaft Torsional Fatigue due to Pressure Spikes: Figure 4.41 shows *shaft* failure was caused by torsional fatigue due to pressure spikes. As shown in the figure, shaft fracture is at random cut.

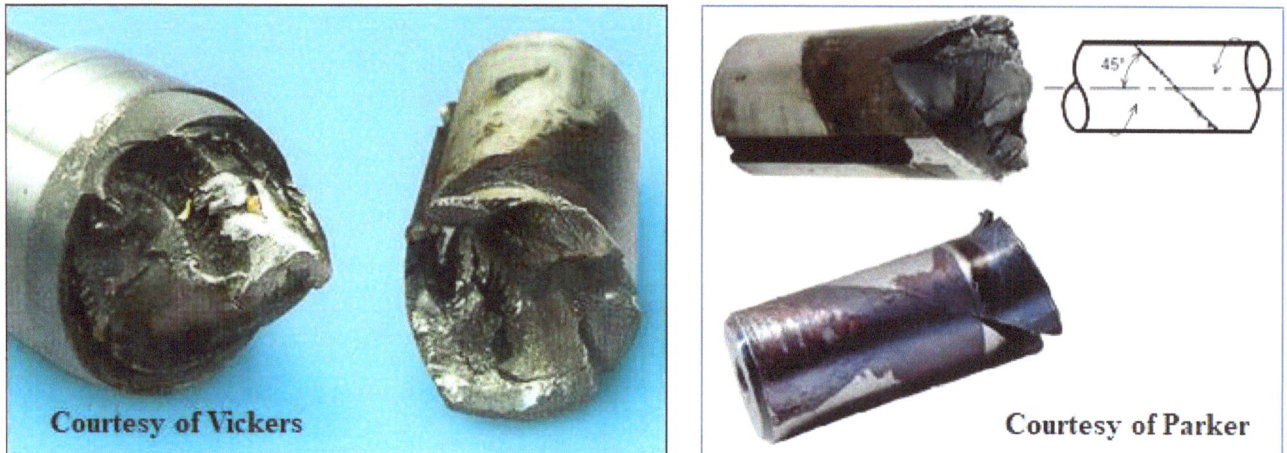

Fig. 4.41- Pump Shaft Torsional Fatigue due to Pressure Spikes

Pump Shaft Bending Fatigue due to Misalignment: Figure 4.42 shows that the *shaft* is broken cleanly at a 90 deg angle to its axis of rotation. This type of failure is due to rotational bending fatigue that makes the shaft flex slightly with each revolution.

Fig. 4.42- Pump Shaft Bending Fatigue due to Misalignment

Pump Shattered Yoke in Piston Pumps: In variable displacement pumps, bolts that secure the pins to the *yoke* subject to continuous loading and unloading forces every time pump is loaded and unloaded. As shown in Fig. 4.43, improper torqueing of one of the bolts, or an unusually high frequency of pump loading/unloading, can cause bolt failure and yoke breakage.

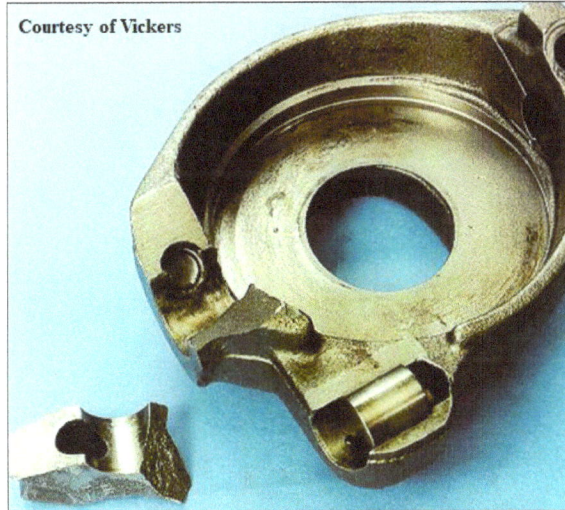

Fig. 4.43- Yoke Fatigue due Frequent Loading

4.3.5- Pump Failures due to Overpressure

Screw Pump: This little screw pump, shown in Fig 4.44, got "screwed" when the outlet pressure exceeded the nominal value by four times. As a result, screws are jammed, and the pump is unexpectedly stopped. Hence, the shaft is twisted, and the key is damaged.

Fig. 4.44- Yoke Fatigue due Frequent Loading

Gear Pump: Pumps operating under a continuous load at high pressure and for extended periods are susceptible to premature wear and failure. Forces generated by the pump outlet pressure and gear area causes a deflection of the gears. This stresses the bushings that support the gear journals. The oil film needed to lubricate and cushion the pump elements becomes thinner with the increase in pressure until direct contact is made. As a result, as shown in Fig. 4.45 (left), pump hosing cutout increases over the time. If the cutouts exceed 0.007" in depth inside the gear case housing, the pump will become less efficient and the housing is to be replaced. The right side of the figure shows excessive pressure and deflection of the gears in the pump housing as a result of loss of the lubricating film between the pump shafts and the bushings. Continued high pressure degrades the bushing and causes the bushing spinning in its housing.

Gear cutout in Pump Housing should not exceed .007"

Fig. 4.45- Example of Gear Pump Failures due to Overpressure (www.alamo-industrial.com)

Vane Pump: Figure 4.46 shows a rotor of vane pump was ruptured (left) and port block was cracked due to overpressure.

Fig. 4.46- Example of Vane Pump Failures due to Overpressure (Courtesy of Parker)

4.3.6- Pump Failures due to Insufficient Charge Pressure

Charge pumps play crucial role in *hydrostatic transmission* systems. Figure 4.47 shows main pump failure caused by insufficient *charge pressure* on startup.

Fig. 4.47- Example of Pump Failure due to Insufficient Charge Pressure on Start-Up

4.3.7- Pump Failures due to Dry Run

As shown in Fig. 4.48, the quantity and dry quality of the chips, bits and scraps, the pump turned for quite a long time without oil.

Fig. 4.48- Example of Pump Failure due to Dry Run (Courtesy of Insane Hydraulics)

4.3.8- Pump Failures due Low Quality

Figure 4.49 shows bronze sleeve somehow detaches itself from the cylinder block. Bad quality? Contamination? Overheating? Could be a combination of all of that.

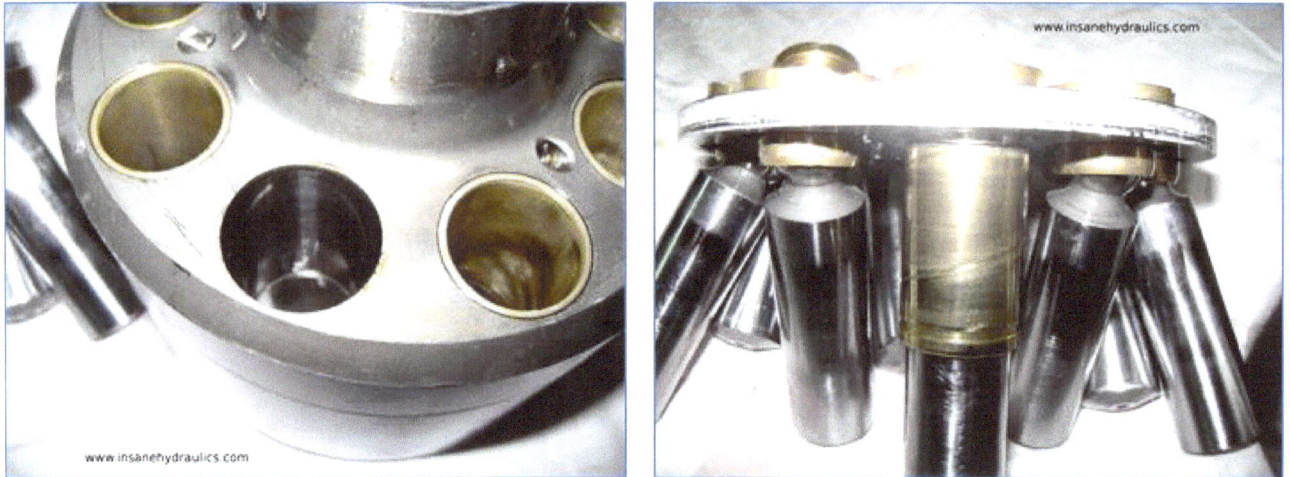

Fig. 4.49- Example of Pump Failure due to Low Quality (Courtesy of Insane Hydraulics)

4.3.9- Pump Failures due Oil Breakdown

The powerpack shown in Fig. 4.50 was functioning for years. Hydraulic fluid didn't change for long time and gradually turning into what is seen in the figure. Although the equipment still functioning and the necessary pressure was still there, the oil temperature got so high that the protective plastic cap on an accumulator charging valve started to literally melt down, before the pump finally gave up. The sludge deposits are the silent but proofs the severe oil over-stress and breakdown.

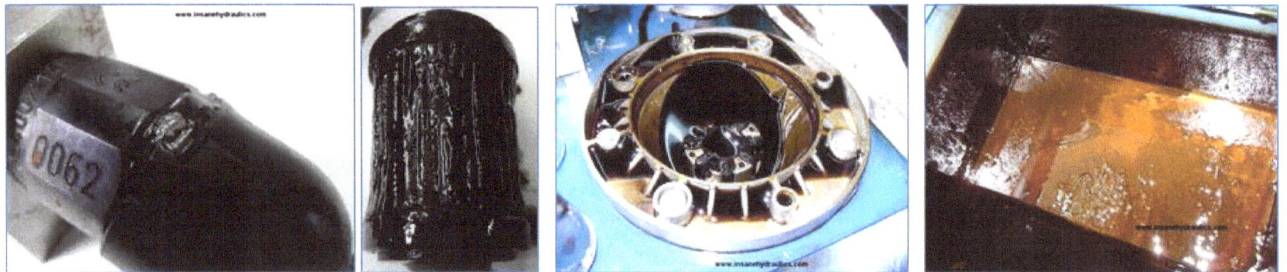

Fig. 4.50- Example of Pump Failure due to Oil Breakdown (Courtesy of Insane Hydraulics)

4.3.10- Pump Failures due Shaft Misalignment

As it seen from the Fig. 4.51, the damage to the cylinder block was not caused by over-torque (the splines are intact), but rather by shaft misalignment, caused by destruction of the tail bearing, and then the front bearing. When you find your machine making strange noises or losing force "just a little", don't let it reach the catastrophic failure point, do something about it, it will be cheaper and cause less downtime in the long run.

Fig. 4.51- Example of Pump Failure due to Shaft Misalignment (Courtesy of Insane Hydraulics)

4.3.11- Pump Failures due Improper Priming

Figure 4.52 shows a pump that worked for a couple of minutes before turning into scrap. This is a perfect example of how a new pump can become old in a very-very short time due to "confident incompetence" of technical personnel involved in hydraulic equipment re-commission. This is a classic case of a bad start-up procedure. It cannot get more classical than this! Obviously, the person responsible for putting the machine back in service didn't have a clue of how that particular circuit was functioning, and assumed that, as long as this type of pump had no drain line and the suction line was internally connected to the casing, thus ensuring all the internal lubrication, there was no need to take any special precautions before starting the engine. Obviously, thus is wrong!!!

First of all, this pump indeed doesn't have a case drain, and the suction line is connected to the case. The pump is designed so that it sucks oil through the openings in the cylinder block, and so the suction flow through the casing is providing all the necessary lubrication and cooling. The leakage goes to casing and is sucked back again by the rotary group. Pretty standard arrangement, eliminating the need for a case drain. So, what went wrong?

Well, one thing wasn't taken into consideration in that particular system, pump isn't primed.

Fig. 4.52- Example of Pump Failure due to Improper Priming (Courtesy of Insane Hydraulics)

4.3.12- Pump Failures due Lack of Overhauling

Something should be naturally expected from a 20,000-hour old rotary group. The oil and the filters were changed at regular intervals, but no scheduled overhauls of the pump were ever performed. The result was, As shown in Fig. 4.53, a relatively long life of the unit, which still ended abruptly and unexpectedly when the parts that wear out, well, wore out.

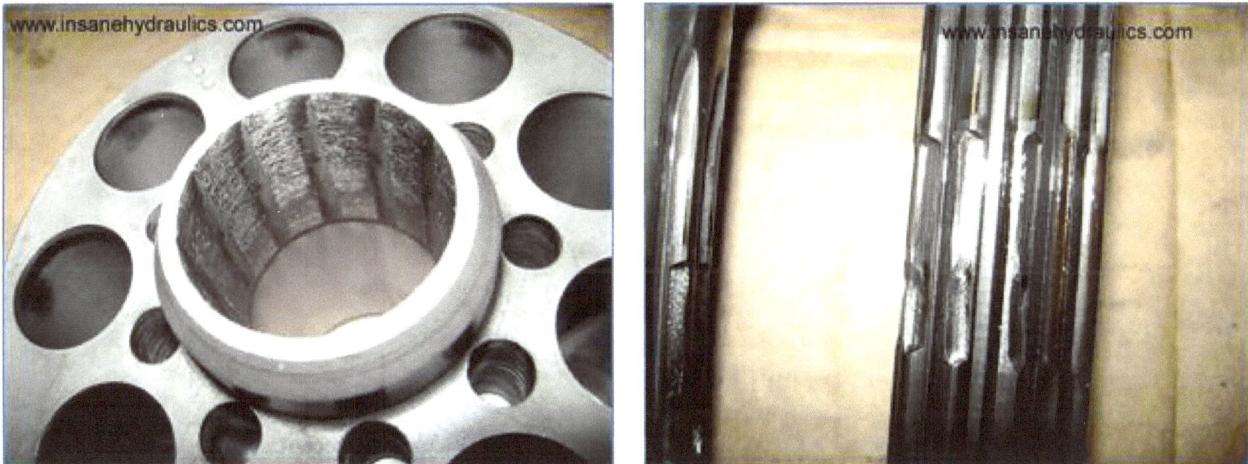

Fig. 4.53- Example of Pump Failure due to Lack of Overhauling (Courtesy of Insane Hydraulics)

Chapter 5

Troubleshooting and Failure Analysis of Motors

Objectives

This chapter discusses hydraulic *motors* inspection, troubleshooting, and failure analysis. In this chapter, a troubleshooting chart for motor faults is presented. The chapter also presents examples of defective motors due to various reasons such as contamination, clogged case drain, shaft failure, etc.

Brief Contents

5.1- Hydraulic Motors Inspection

5.2- Hydraulic Motors Troubleshooting

5.3- Hydraulic Motors Failure Analysis

Chapter 5: Troubleshooting and Failure Analysis of Motors

5.1- Hydraulic Motors Inspection

Hydraulic *motors* have wide varieties based on motor displacement (fixed or variable), rotation (unidirectional or bidirectional), and motor mechanism (gear, vane, piston, and High-Torque Low-Speed "HTLS"). Volume 1 of this series of textbooks presents good idea about the construction and operating principle of various motor mechanisms. Table 5.1 shows a typical inspection sheet for a hydraulic motor.

Hydraulic Motor Inspection Sheet	
Manufacturer	
Model #	
Serial #	
Location	
Motor Mechanism	☐ External Gear ☐ Internal Gear ☐ Gerotor ☐ Vane Motor ☐ Radial Piston [☐ Rotating Cam ☐ Rotating Cylinder Block] ☐ Bent Axis ☐ Swash Plate ☐ Other [ ]
Direction of Rotation	☐ Unidirectional ☐ Bidirectional
High Torque Low Speed Motor (HTLS)	☐ Yes ☐ NO
Motor Displacement	☐ Fixed ☐ Variable [= cc/rev]
Type of Control	☐ Pressure Compensated ☐ Displacement Controlled ☐ Constant Power (Torque) ☐ Load Sense
Drive Shaft	Type and Size:
Ports	Case Drain: ☐ Yes ☐ NO Case Drain size: Inlet Port size: Outlet Port Size:
Conditions of Seals	
Conditions of Bearings	
Conditions of Inside Parts	
Other Notes:	

Table 5.1 – Hydraulic Motors Inspection Sheet

5.2- Hydraulic Motors Troubleshooting

Table 5.2 shows the troubleshooting chart for hydraulic motors. Many of the faults of motor mechanisms are similar to what occurs in pump mechanisms. Therefore, some of the pump troubleshooting charts will be used in motor troubleshooting. So, when any of the pump troubleshooting charts are indicated here, **it should be used as applicable for motors.**

T-Motor-01-Motor Troubleshooting	
Leaking motor?	▪ Consult Chart (as applicable for motors): ▪ **"T-Pump-09-Leaking Pump"**.
Excessive motor wear?	▪ Consult Chart (as applicable for motors): ▪ **"T-Pump-10-Excessive Pump Wear"**.
Excessive motor noise and vibration?	▪ Consult Chart (as applicable for motors): ▪ **"T-Pump-12-Excessive Pump Noise & Vibration"**.
Motor shows slow performance?	▪ Check flow received by the motor. ▪ Check controller setting for variable motors. ▪ Consult Chart: ▪ **"T-System-10-Actuator Slow Performance"**.
Motor shows fast performance?	▪ Check flow received by the motor. ▪ Check controller setting for variable motors. ▪ Consult Chart: ▪ **"T-System-11-Actuator Fast Performance"**.
Motor shows erratic performance?	▪ Check controller setting for variable motors. ▪ Check if the motor rotates below recommended minimum speed **(Note 1).** ▪ Consult Charts: ▪ **"T-Pump-03-Erratic Flow out of the Pump"**. ▪ **"T-System-12-Actuator Erratic Performance"**.
Motor rotates in wrong direction?	▪ Consult Chart: ▪ **"T-System-13-Actuator Moves in Wrong Direction"**.
Motor fails to rotate?	▪ Consult Chart: ▪ **"T-System-14-Actuator Stops to Move"**.

| Motor load drifts? | ▪ Consult Chart: **"T-System-15-Actuator Load Drifts"**. |
| Motor leaks? | ▪ Consult Chart: **"T-System-16-Actuator Leaks"**. |

Table 5.2 – Hydraulic Motors Troubleshooting Chart

Note 1: Figure 5.1 shows motor speed n_m as function of the motor flow Q_m and motor displacement D_m. Obviously, n_m increases with the decreasing of D_m and/or increasing Q_m; and vise versa. If the motor is fully de-stroked, its speed will go theoretically to infinity. Practically in this case, motor speed and internal friction increases to the limit that the mechanical efficiency drastically dropped, the whole input power is wasted as heat, and no torque is produced on the motor shaft. Therefore, minimum motor displacement must comply with the maximum speed and allowable minimum mechanical efficiency. Motor manufacturers usually add a mechanical stopper to limit the minimum motor displacement. Increasing motor displacement will slow down the motor. Maximum motor displacement should comply with the minimum speed of the motor, below which the motor may run erratically. It must also comply with the maximum torque of the motor. Both motor minimum and maximum speeds of a motor are recommended by the motor manufacturer.

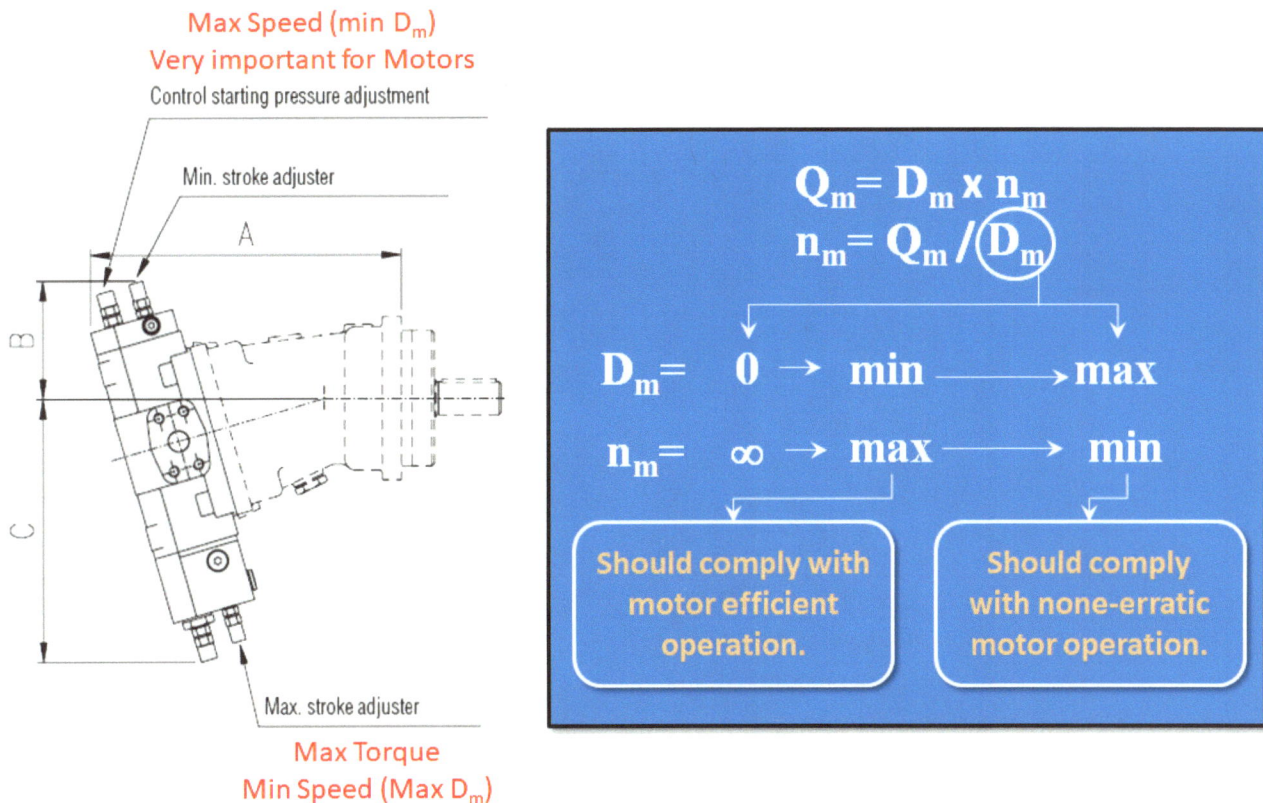

Fig. 5.1 – Recommended Speed Range of Hydraulic Motors

5.3- Hydraulic Motors Failure Analysis

A final drive or a hydraulic motor represents a major investment in any equipment, not just because of the cost, but because hydraulic motors keep the fleet running. Whether a full-size excavator with a track drive motor, a compact track loader with a final drive motor, or a skid steer loader with a propel motor. When any hydraulic motor fails, the rest of the machine stops.

As it has been previously stated, many of the faults of motor mechanisms are like what occurs in pump mechanisms. Therefore, symptoms of motors failures are similar to that have been stated in pump failure analysis. However, the following examples are specific to motors.

Final Drive Motor Failure due to Clogged Case Drain Filter: Severe damage can occur if the case drain filter isn't checked on a regular basis because it will become clogged. As it clogs, pressure will build because hydraulic fluid can no longer freely pass through the case drain line. Pressure will continue to build, eventually forcing hydraulic fluid into the gear hub, blowing seals, piston shoes, and bearings along the way. If case drain pressure keeps building up, it can cause damage as it seen in Fig. 5.2 for a swash plate axial piston motor.

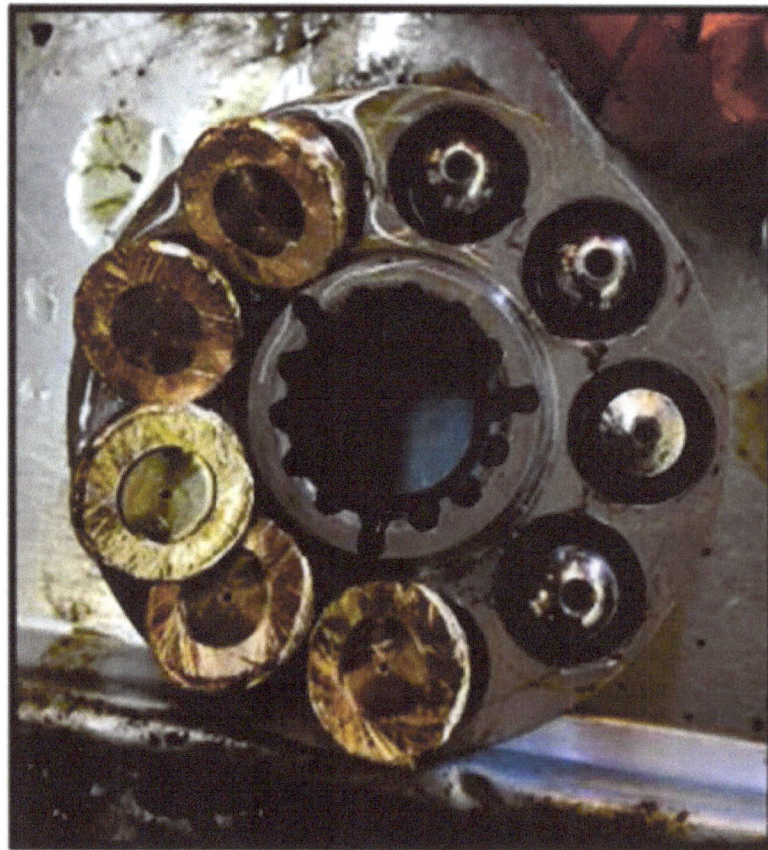

Fig. 5.2 – Example of Hydraulic Motor Failure due to Clogged Case Drain Filter (shop.finaldriveparts.com)

Final Drive Motor Failure due to Abrasive Contaminants: Figure 5.3 shows a failed radial piston motor. Abrasive particles can scratch sensitive surfaces, reduce the effectiveness of bearings, damage gear teeth, scar pistons, drastically reduce efficiency, and more. If the contamination isn't addressed, it will lead to expensive and irreparable damage.

**Fig. 5.3 – Example of Hydraulic Motor Failure due to Contamination
(http://info.texasfinaldrive.com/)**

Hydraulic Motor Shaft Failure due to Overload: *Shock* loads and sudden *pressure spikes* that exceed the pressure rating of the pump can cause pump and motor shaft failure. Examples of that is when is when a forestry machine moving fast while cutting heavy brush or trees. Figure 5.4 shows two different modes of pump shaft failure due to shock load as follows:
1. Transverse torsional shear due to a high, single overload application.
2. Torsional overload resulting in transverse shear with swirl pattern.

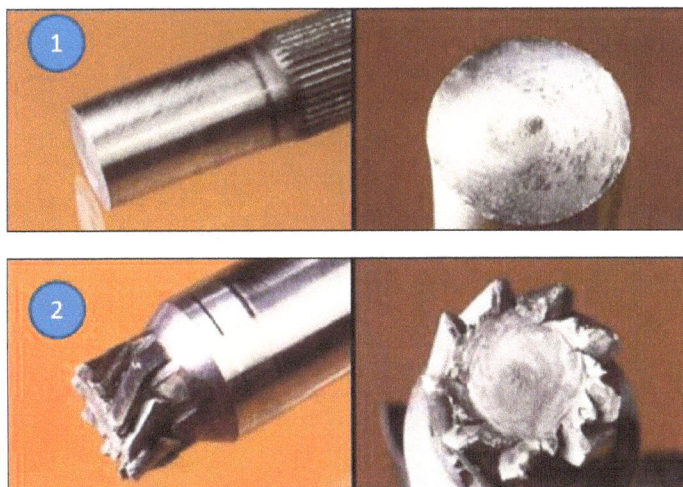

**Fig. 5.4 – Example of Hydraulic Motor Shaft Failure due to Overload
(www.alamo-industrial.com)**

Gear Motor Failure due to Over-torqueing: The motor was working Ok in a *verge shredder*. As shown in Fig 5.5, inspecting the motor shows the following:

- Leak in the shaft seal area.
- front flange was "slightly worn".
- The spline coupling that nice crack in it (1),
- The shaft key was sheared (2). There was essentially no shaft key, the torque transmission between them was secured by the "friction weld" regime.

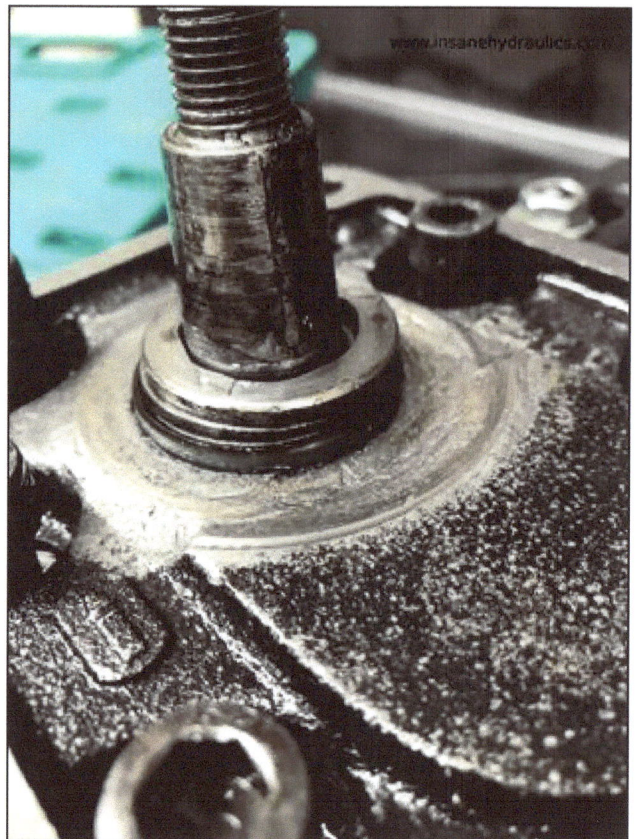

**Fig. 5.5 – Example of Hydraulic Motor Failure due to Overload
(Courtesy of Insane Hydraulics)**

Piston Motor Failure due to Contamination: This is another fixed displacement hydraulic motor that won't be displacing much within a hydrostatic transmission. Oil contamination resulted in the rotary group was completely destroyed as shown in Fig 5.6.

**Fig. 5.6 – Example of Hydraulic Motor Failure due to Oil Contamination
(Courtesy of Insane Hydraulics)**

Chapter 6

Troubleshooting and Failure Analysis of Cylinders

Objectives

This chapter discusses hydraulic *cylinders* inspection, troubleshooting, and failure analysis. In this chapter, a troubleshooting chart for cylinder faults is presented. The chapter also presents examples of defective cylinder due to various reasons such as contamination, improper mounting, improper load attachment, side loading, overpressure, overheating, fluid incompatibility, saltwater, external leakage, etc.

Brief Contents

6.1- Hydraulic Cylinders Inspection

6.2- Hydraulic Cylinders Troubleshooting

6.3- Hydraulic Cylinders Failure Analysis

Chapter 6: Troubleshooting and Failure Analysis of Cylinders

6.1- Hydraulic Cylinders Inspection

Hydraulic *cylinders* could be of a tie-rod or mill type configuration. Volume 1 of this series of textbooks presents good idea about the construction and operating principles of hydraulic cylinders. Table 6.1 shows a typical inspection sheet for a hydraulic cylinder.

Hydraulic Cylinder Inspection Sheet	
Manufacturer	
Model #	
Serial #	
Location	
Cylinder Configuration	☐ Single Acting [☐Spring Return ☐ Load Return] ☐ Double Acting Differential [☐Single Rod ☐ Double Rods] ☐ Double Acting Synchronous ☐ Telescopic [☐ Single Acting ☐ Double Acting] Body [☐ Tie Rod ☐ Mil Type] Cushions [☐ Yes ☐ No]
Dimensions	☐ Bore ( ) ☐ Rod ( ) ☐ Stroke ( )
Ports	Port size:
Operation	☐ Push ☐ Pull ☐ both
Mounting Type	☐ Front Flange ☐ Rear Flange ☐ Pin Eye ☐ Trunnion
Position	☐ Vertical ☐ Horizontal ☐ Angle
Conditions of Rod & Load attachment	
Conditions of Rod Seal	
Conditions of Piston	
Conditions of Piston Seal	
Condition of Barrel Inside surface	
Conditions of Cyl. Cap	
Conditions of Cyl. Head	
Conditions of Ports	
Other Notes:	

Table 6.1 – Hydraulic Cylinders Inspection Sheet

6.2- Hydraulic Cylinders Troubleshooting

Table 6.2 shows the troubleshooting chart for hydraulic cylinders.

T-Cylinder-01: Cylinder Troubleshooting	
External leakage from rod seals?	▪ Check cylinder rod for scoring, galling, and/or bending. ▪ Check fluid cleanliness. ▪ Consult Chart: **"T-Seal-01-Seal Troubleshooting"**.
External leakage from between the barrel and the end caps?	▪ Pressure too high. ▪ Check the tie rod torque. ▪ Check/replace static seal between barrel and end caps. ▪ Consult Chart: **"T-Seal-01-Seal Troubleshooting"**.
Cylinder internal leaking?	▪ Check if the cylinder is over pressurized. ▪ Check piston seal deterioration. ▪ Consult Chart: **"T-Seal-01-Seal Troubleshooting"**.
Scored cylinder rod?	▪ Replace/cover the rod **(See Note 1)**.
Bent cylinder rod? Bushing wear? Galling piton rod?	▪ Check load misalignment with the load. ▪ Check maximum allowable bending and replace the rod if needed **(See Note 2)**. ▪ Check side loads on the rod and consider using stop tubes if needed **(See Note 3)**. ▪ Check cylinder buckling due to exceeding maximum compressive load **(See Note 4)**. ▪ Check improper mounting to the machine body or load attachment.
Cylinder shows slow performance?	▪ Check flow received by the cylinder. ▪ Consult Chart: ▪ **"T-System-10-Actuator Slow Performance"**.
Cylinder shows fast performance?	▪ Check flow received by the cylinder. ▪ Consult Chart: ▪ **"T-System-11-Actuator Fast Performance"**.
Cylinder shows erratic performance?	▪ Consult Charts: ▪ **"T-Pump-03-Erratic Flow out of the Pump"**. ▪ **"T-System-12-Actuator Erratic Performance"**.
Cylinder moves in wrong direction?	▪ Consult Chart: ▪ **"T-System-13-Actuator Moves in Wrong Direction"**.
Cylinder stops to move?	▪ Consult Chart: **"T-System-14-Actuator Stops to Move"**.
Hydraulic cylinder has no cushioning effect/moves hard into the end position?	▪ The end position cushioning setting does not comply with the requirements

Table 6.2 – Hydraulic Cylinders Troubleshooting Chart

Note 1 (Covering the Rod): In some cases, as shown in Fig. 6.1, cylinders may work in harsh environments such as nearby welding area or in a highly contaminated and dusty environment. In such cases, cylinder wipers can't cope with this level of *contamination*. Therefore, it is highly recommended to cover the cylinder rod.

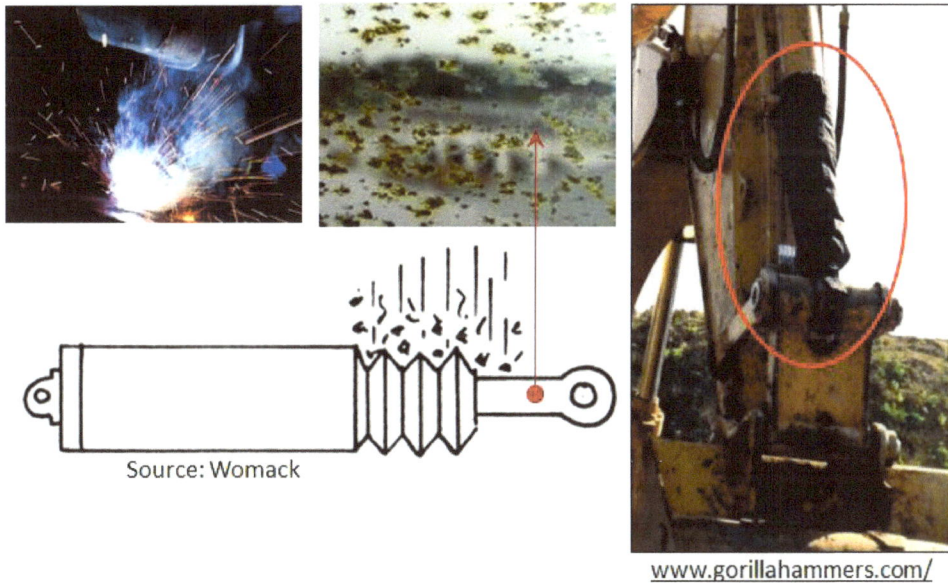

Source: Womack

www.gorillahammers.com/

Fig. 6.1 – Protecting Cylinder Rod from Harsh Environment

Note 2 (Check for Rod Bending): As shown in Fig. 6.2, cylinder rod must be checked for *straightness*. Based on the cylinder length there is a maximum allowable bend value (review Volume 5 of this series of textbooks).

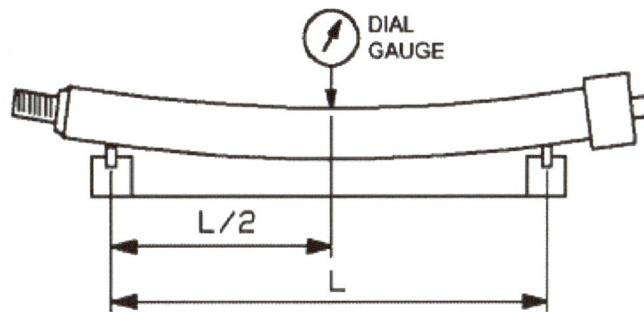

Fig. 6.2 – Checking Rod Straightness

Note 3 (Using Stop Tube): As shown in Fig. 6.3, the cylinder rod is a third-class lever with the piston acting as a fulcrum at one end, and the side load (side force) acting on the rod at the other end. The side load acting on the rod end has a tendency to rotate the rod about the piston. Overloading the cylinder in this manner will wear the rod bearing and rod seals on one side, and the piston and the cylinder bore on the other side. If the lubrication film fails, rod and bushing galling can be expected. Use of *Stop Tube* reduce the reaction and protect cylinder bushing and seals.

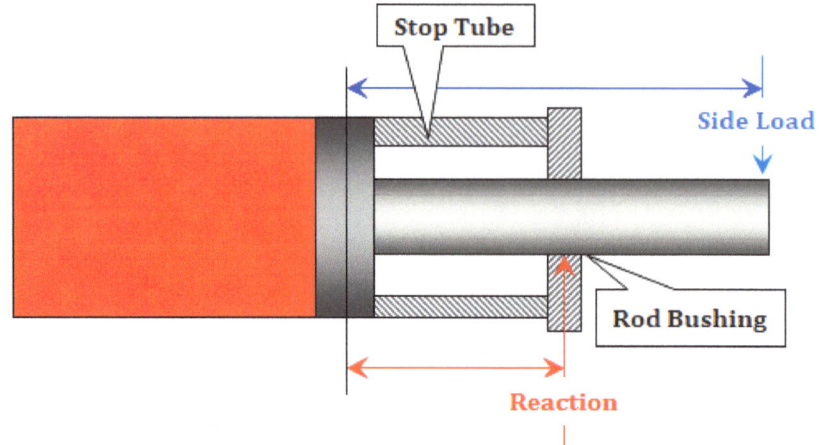

Fig. 6.3 – Use of Stop Tube to Protect Rod Bushing

Note 4 (Check for Buckling): As shown in Fig. 6.4, cylinder rod buckling is factor of cylinder dimensions, material, compressive load, and the way the cylinder is attached to the load. Cylinder rod buckling destroys the rod bushing and seals causing external leakage.

Fig. 6.4 – Cylinder Rod Buckling

6.3- Hydraulic Cylinders Failure Analysis

6.3.1- Cylinder Failure due to Particulate Contamination

Common Worn Areas due to Contamination: Figure 6.5 shows the opposing moving surfaces within hydraulic cylinders that are commonly affected by abrasives.

- **At Rod Seals:** The clearance between the cylinder rod, rod seals and wipers.
- **At Piston Seals:** The clearance between the piston seal package and the cylinder barrel.

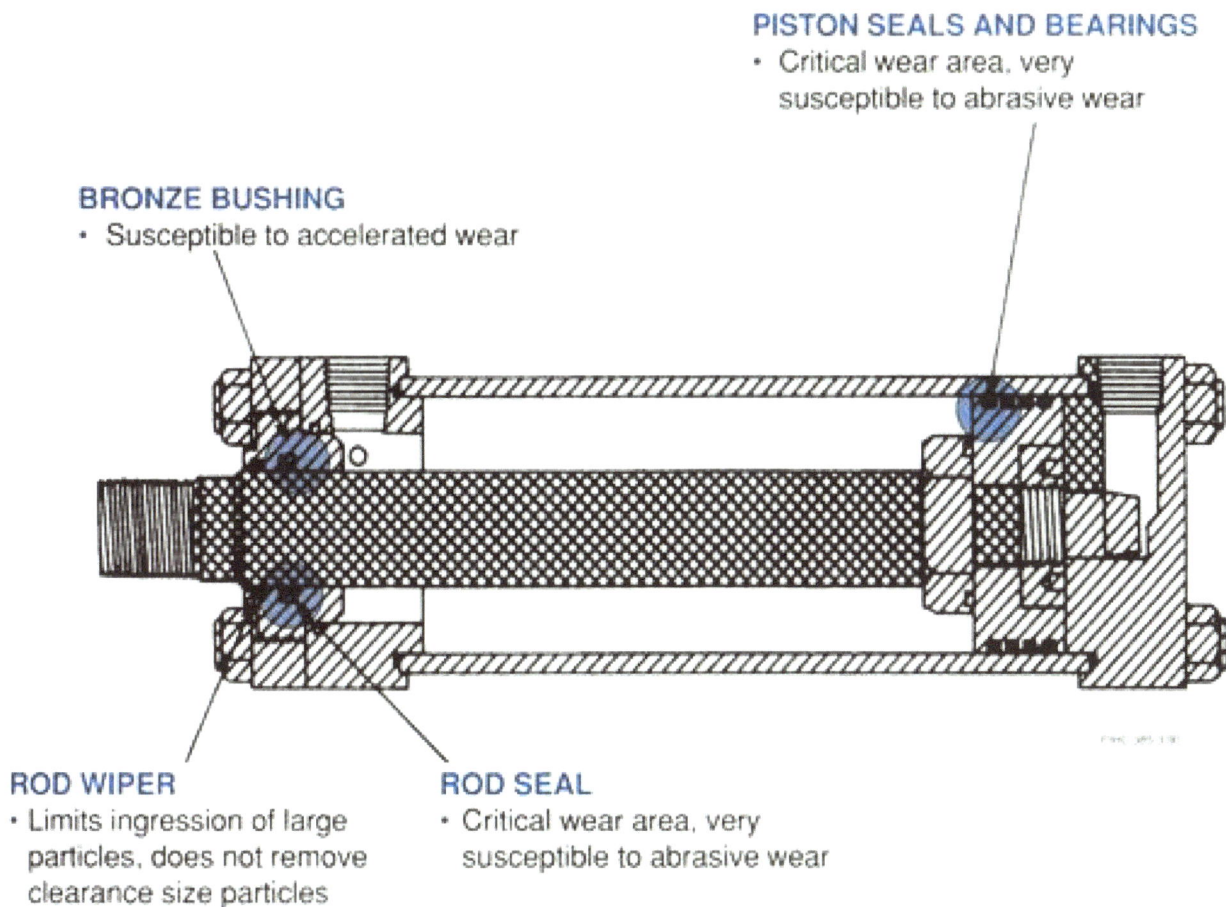

PISTON SEALS AND BEARINGS
- Critical wear area, very susceptible to abrasive wear

BRONZE BUSHING
- Susceptible to accelerated wear

ROD WIPER
- Limits ingression of large particles, does not remove clearance size particles

ROD SEAL
- Critical wear area, very susceptible to abrasive wear

Fig. 6.5- Commonly Worn Areas within Hydraulic Cylinders (Courtesy of Pall)

Cylinder Failures due to Contamination: Figure 6.6 shows examples of hydraulic cylinder failure due to particulate contamination. The upper part of the figure shows visible leakage due to seal failure caused by abrasive particulate contamination. The figure shows (on lower left) piston rings that were eaten away by contaminants. On lower right side, the figure shows a scored cushion bushing resulting in a loss of cushioning effect.

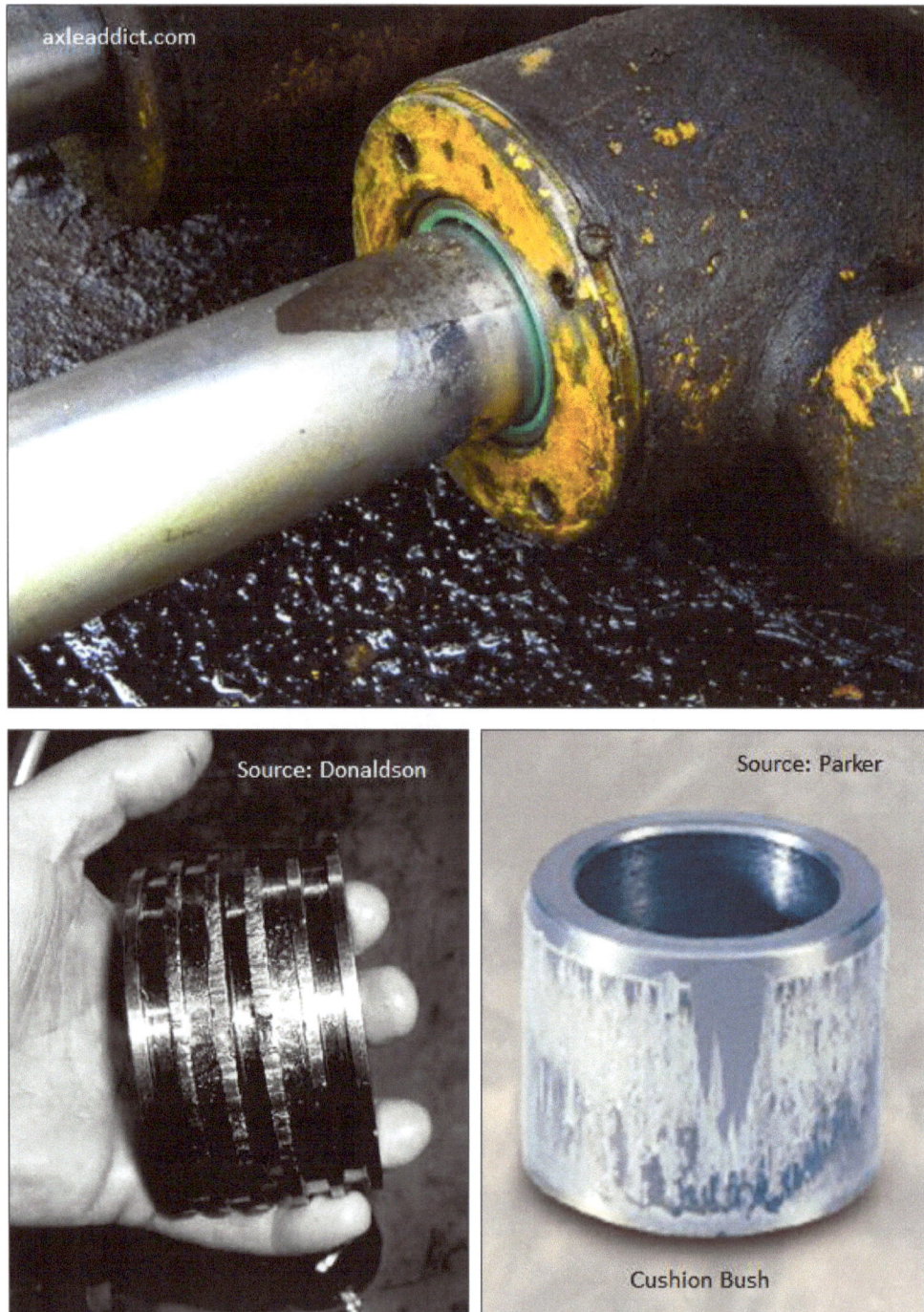

Fig. 6.6- Examples of Hydraulic Cylinder Failures due to Particulate Contamination

6.3.2- Cylinder Failure due to Improper Mounting

Cylinder Failures due to Improper Mounting: Hydraulic cylinders should be properly mounted on the machine body and properly attached to the load (in this regard, review volume 5 of this series of textbooks). Otherwise, as shown in Fig. 6.7, rod *bending*, and structural damage could occur.

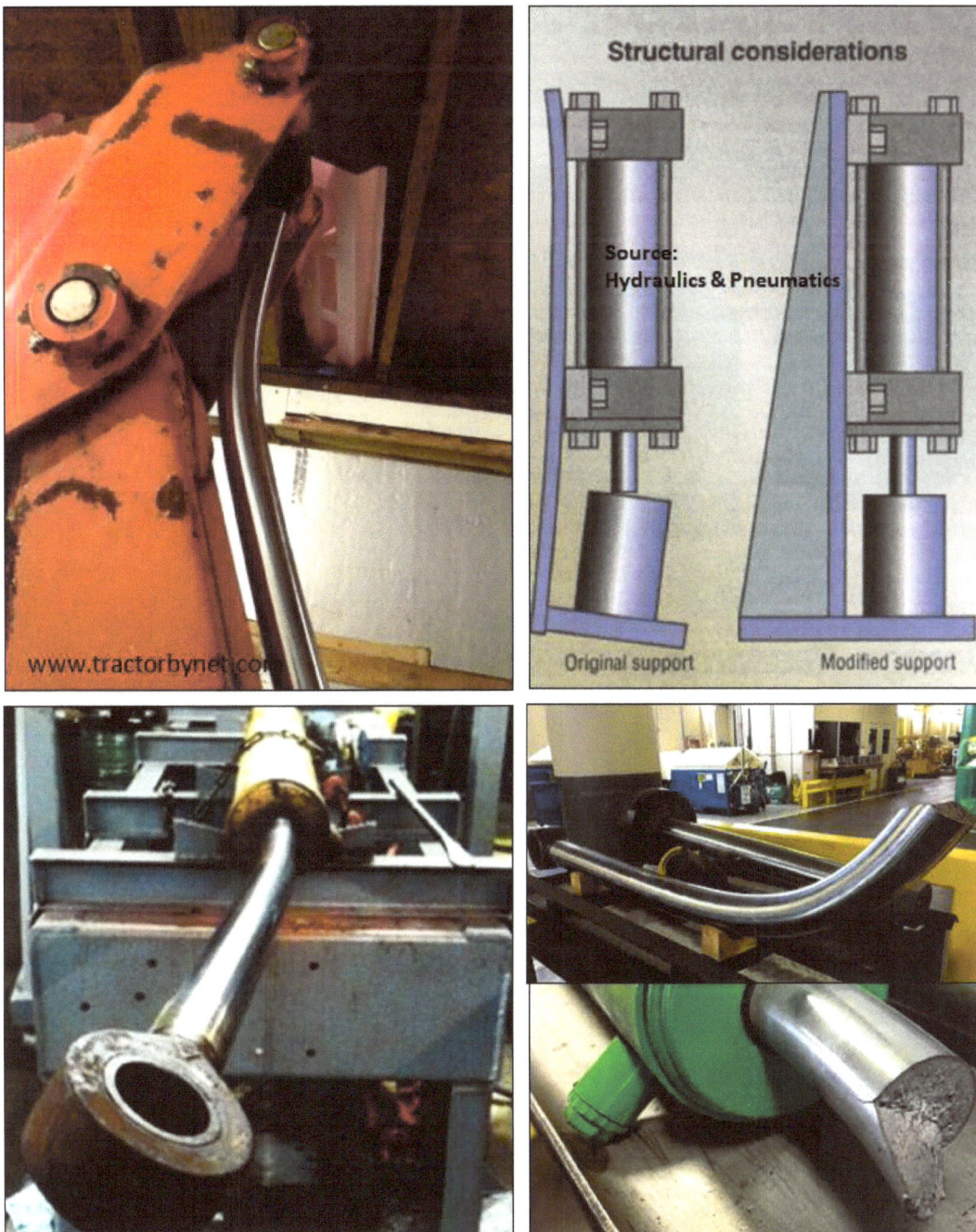

Fig. 6.7- Examples of Hydraulic Cylinder Failures due to Improper Mounting

6.3.3- Structural Failure due to Improper Load Attachment

Structural Failure due to Improper Load Attachment: As shown in Fig. 6.8, the aerial work platform collapsed during the incident and the lift basket crashed to the ground and landed on the grassy surface by the Field House (Photo courtesy of OSHA). The collapse of the aerial work platform was determined to be caused by a failure of the upright level cylinder. The rod of the cylinder had broken away from the cylinder barrel. Investigation on the case showed that the cylinder had been modified and the modification directly caused the failure of the cylinder. The original rod assembly of the cylinder had a threaded end. The piston was secured to the rod by a hex nut that screwed into the threaded end. The hex nut had a set screw to secure it to the rod. The investigation found that, as shown in the figure, the thread of the rod assembly of the failed cylinder was ground off. A hole was drilled through the hex nut and the rod. A bolt had been inserted through the hex nut and the rod and it was fastened with a nut. The bolt broke into three pieces during the incident and caused the cylinder to fail

The thread of the failed cylinder rod was ground off and a hole had been drilled through the rod.

The bolt broke into three pieces during the incident and caused the cylinder to fail.

The hex nut was drilled through so that a bolt can be inserted to hold the rod and the nut together

**Fig. 6.8- Example of Structural Failure due to Improper Load Attachment
(metroforensics.blogspot.com)**

6.3.4- Cylinder Failure due to Side Loading

Cylinder Rod and Busing Galling due to Side Loads: Figure 6.9 shows *galled* cylinder rod due to side loading. *Gland* and seal showing wear resulting from excessive *side loading*. Brass on the seal is a mixture of gland and seal material.

Fig. 6.9- Examples of Rod, Bushing, and Seal Failure due to Side Loads (Courtesy of Parker)

Cylinder Rod Bending due to Side Loads: Figure 6.10 shows harsh bending of a cylinder rod due to excessive side loading.

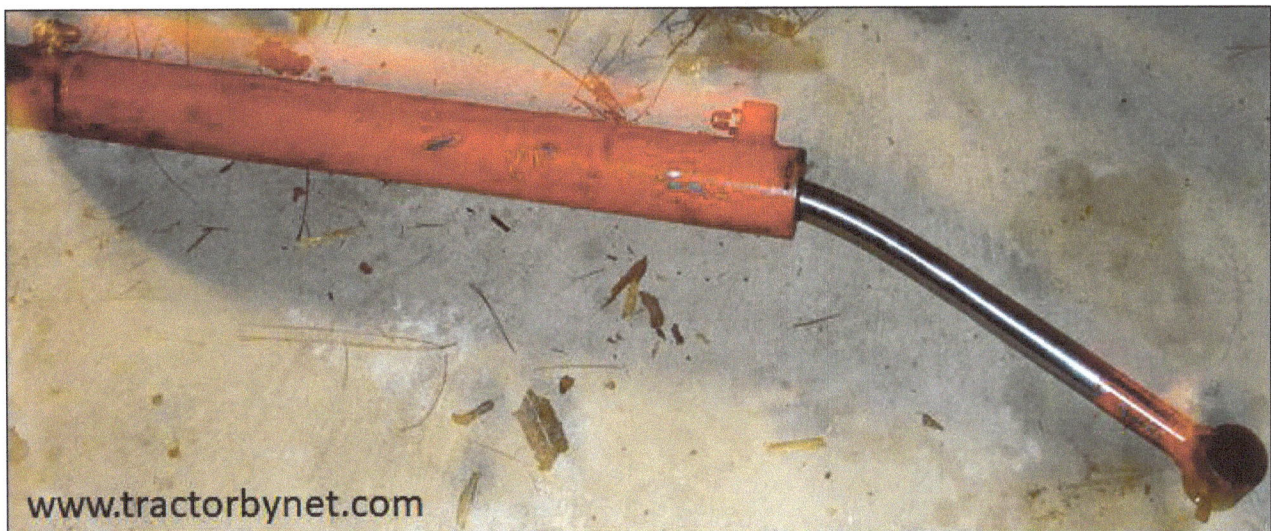

Fig. 6.10- Examples of Cylinder Rod Bending due to Side Loading

6.3.5- Cylinder Failures due to Over Pressurization

Cylinder Burst and Seal Extrusion due to Excessive Working Pressure: When working pressure increases uncontrollably, cylinder burst could occur as shown in Fig. 6.11. The figure shows wiper seal extrusion damage as a result of overpressure.

Fig. 6.11- Examples of Cylinder Rod Bending due to Overpressure

6.3.6- Cylinder Seal Failures due to Over Heating

Cylinder Seal Failure due to Excessive Working Temperature: Figure 6.12 shows piston seal softness (left) and cracked wiper lip (right) caused by excessive temperature.

Fig. 6.12- Examples of Piston Seal and Rod Wiper Failure due to Overheating

6.3.7- Cylinder Seal Failures due to Fluid Incompatibility

Cylinder Seal Failure due to Fluid Incompatibility: Figure 6.13 shows a new piston on the left side. The seal on right is damaged by water, a common contaminant in mineral oil system.

Fig. 6.13- Examples of Piston Seal Damage due to Fluid Incompatibility

6.3.8- Cylinder Rod Corrosion due to Saltwater

As shown in Fig. 6.14, as a result of using low quality stainless-steel material used in making cylinder rods, pitting on cylinders rods is typically caused by corrosion from saltwater in marine applications. The lip seals on a shaft need a smooth surface or the hydraulic fluids will leak out into the ocean. Such rods are unrepeatable and must be scrapped.

Fig. 6.14- Example of Cylinder Rod Corrosion due to Saltwater

6.3.9- Cylinder External Leakage

Cylinder External Leakage: Figure 6.15 shows small leakage (upper figures) that accumulates over the time around the cylinder head. In some cases (lower figure) leakage may be excessive in continuous stream.

Fig. 6.15- Examples of Cylinder External Leakage

6.3.10- Cylinder Rod Collapse due to Pressure Intensification

Some hydraulic cylinders use hollow rods. When the rod diameter is large relative to the piston diameter, that results in a significantly high differential area ratio and the cylinder becomes a very efficient pressure intensifier. As a result, when meter out flow control or the return line from the rod end is restricted during extension, cylinder rod can collapse as shown in Fig. 6.16.

Fig. 6.16- Examples of Cylinder Rod Collapse due to Pressure Intensification

Chapter 7

Troubleshooting and Failure Analysis of Valves

Objectives

This chapter discusses hydraulic *valves* inspection, troubleshooting, and failure analysis. In this chapter, a troubleshooting chart for valve faults is presented. The chapter also presents examples of defective hydromechanical and electrohydraulic valves due to various reasons such as particulate and chemical contamination, solenoid burning due to inrush current, etc.

Brief Contents

7.1- Hydraulic Valves Inspection

7.2- Hydraulic Valves Troubleshooting

7.3- Hydraulic Valves Failure Analysis

Chapter 7: Troubleshooting and Failure Analysis of Valves

7.1- Hydraulic Valves Inspection

Hydraulic *valves* have various operating characteristics based on valve control function (pressure, flow, and directional), valve operation (direct or pilot), and control (direct or pilot). Volume 1 of this series of textbooks presents the construction and operating principles of various valves. Table 7.1 shows typical inspection sheet for a hydraulic valve.

Hydraulic Valve Inspection Sheet	
Manufacturer	
Model #	
Serial #	
Location	
Pressure Control Valve Type	• Direct [☐Relief ☐Counterbalance ☐Sequence ☐Reducing] • Pilot [☐Unloading ☐Over-Center ☐ Motor Brake]
Directional Control Valve Type	• # Ports ( ) # Positions ( ) • Initial/Central Position: ( ) • Reset [☐Spring ☐Detent] • Actuation: [☐Manual ☐Mechanical ☐Pilot ☐Electrical] • More info ( )
Flow Control Valve Type	☐Throttle ☐Regulator
EH Valve	Type: [☐ON/OFF ☐Proportional ☐Servo] Signal: ( ) Current = Voltage = Power:
Valve Configuration	Operation: [☐Direct "Single-Stage" ☐ Pilot "Multiple stages"] Control: [☐Direct "Internal" ☐ Pilot "External"] Drain: [☐ Internal ☐ External] Built-in Check Valve [☐ Yes ☐ No]
Moving Element:	☐ Poppet Type ☐ Spool Type [☐Linear ☐Rotary]
Mounting	☐ Subplate ☐ Line ☐ Manifold "Screw-In" ☐ Sandwich ☐ Other:
Ports/Flow	Port size = Rated flow Rate =
Conditions	Parts: Seals:
Other Notes:	

Table 7.1 – Hydraulic Valves Inspection Sheet

7.2- Hydraulic Valves Troubleshooting

The first two actions in every valve type troubleshooting chart are common, that is why they are highlighted by blue background. Tables 7.2 through 7.4 are for troubleshooting DCV, FCV, and PCV in sequence.

T-Valve-01: DCV Troubleshooting	
General Inspection	▪ Consult Chart: ▪ **"T-Valve-05-General Valve Troubleshooting"**.
Valve works electro-hydraulically?	▪ Consult Chart: ▪ **"T-Valve-04-EH Valve Troubleshooting"**.
Improper directional control operation?	▪ Incorrect assembly of non-identical valve spool or using the wrong spool? ▪ Review the valve data sheet and ordering code. ▪ Sticking valve spool
Load drifts?	▪ Valve null position miss-adjustment? ▪ Follow the manufacturer's instructions about valve null adjustment.

Table 7.2– DCV Troubleshooting Chart

T-Valve-02: FCV Troubleshooting	
General Inspection	▪ Consult Chart: ▪ **"T-Valve-05-General Valve Troubleshooting"**.
Valve works electro-hydraulically?	▪ Consult Chart: ▪ **"T-Valve-04-EH Valve Troubleshooting"**.
Improper flow control Operation?	▪ Flow control valve may be installed backward. ▪ Check valve size and setting. ▪ Check pressure before and after the valve. ▪ Plugged orifice. ▪ Flow regulator compensator spool sticking. ▪ Check FCV direction **(Note 1)**.
Fluid viscosity too low or too high?	▪ Check the recommended range of working temperature for the valve.

Table 7.3– FCV Troubleshooting Chart

Note 1: Proper installation of flow control valve may be difficult to detect. If a flow control valve is installed backward, it changes the actuator motion. It also affects the type of speed control. As shown in Fig. 7.1, as a result of installing the FCV backward, meter-in control for extension stroke (left) becomes meter-out control of retraction stroke (right). Situation becomes even worse if the valve is installed on the rod side of the cylinder. If it was originally meter-in control and mistakenly the valve installed backward, speed control will become meter-out and pressure is intensified at that side to a limit that may damage this line.

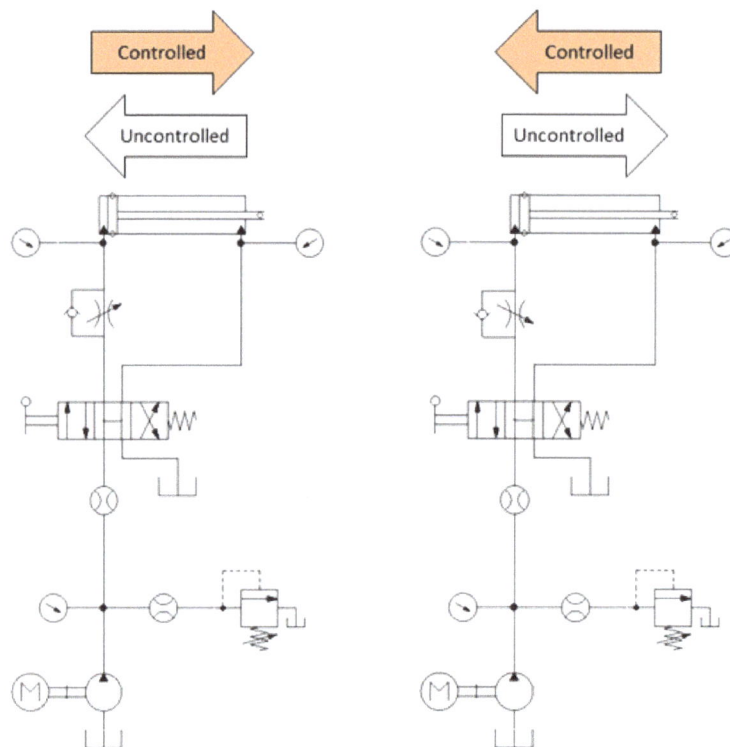

Fig. 7.1 – Backward Installation of a Flow Control Valve Change the Type of Control

T-Valve-03: PCV Troubleshooting	
General Inspection	▪ Consult Chart: ▪ "T-Valve-05-General Valve Troubleshooting".
Valve works electro-hydraulically?	▪ Consult Chart: ▪ "T-Valve-04-EH Valve Troubleshooting".
Improper pressure control Operation?	▪ Check valve size and setting. ▪ Check for worn or broken internal parts.

Table 7.4 – PCV Troubleshooting Chart

In addition to the main control function, if the valve is actuated electrically, troubleshooting chart shown in Table 7.5 is applied. Before troubleshooting EH valve obtain valve documents.

T-Valve-04: EH Valve Troubleshooting	
Valve spool isn't moving?	▪ Use manual override **(See Notes 1)** to check whether it is a mechanical or electrical problem. ▪ Hydraulic: ▪ Valve spool is seized due to heavy *contamination*. ▪ Valve *drain* is blocked. ▪ Pilot or main spool aren't selected properly. ▪ Mechanical: ▪ Valve body is harshly distorted. ▪ Electrical: ▪ No electrical signal is received. ▪ Solenoid is burned due to low voltage, spool seizure, or high flow forces.
Valve spool response is sluggish?	▪ Hydraulic: ▪ Valve is contaminated. ▪ Valve drain is restricted. ▪ Valve body is distorted. ▪ Low pilot pressure. ▪ Electrical: ▪ Supply voltage is too low. ▪ Check setting of *Dither* signal or proportional gain. ▪ Valve is so hot so that EMF is affected **(See Notes 2)**.
Electrohydraulic valve shuddering?	▪ Defective solenoid or *voltage* too low. ▪ Valve is undersized for the flow. ▪ Improper control settings or signal. ▪ Valve cables unshielded and affected by noise. ▪ Valve is cycled too fast **(See Notes 3)**.
External or internal electrical short?	▪ Check wiring conditions and make sure to remove reasons for possible short circuits.
Valve isn't receiving electrical control signal?	▪ Check for broken control signal wire. ▪ Check the signal generator or the controller. ▪ Check fuses or control circuit problem.
ECU of the valve isn't receiving feedback signal?	▪ Check for broken feedback signal wire. ▪ Check defective transducer, limit switch, etc.
Voltage too high or too low?	▪ Check the power supply condition vs. the nominal ratings **(See Notes 4)**.
Solenoid or torque motor of an EH valve is burned?	▪ Overlap energizing two solenoids **(See Notes 5)**. ▪ Contaminated valve cause valve seizure and burning coils due to inrush current.

Table 7.5 – EH Troubleshooting Chart

Note 1 (Manual Override): Optionally, EH valves can be equipped with *Manual Override* feature. This means that, as shown in Fig. 7.2, a little push pin is integrated to the armature so that the armature can be manually forced to move in cases like the following:

- During emergency situations (e.g. power supply failure): the valve spool can be manually shifted to complete the cycle or to release pressure in the valve.
- During system troubleshooting: the valve spool can be shifted manually to check if the problem is from the electrical signal or from a mechanical failure.

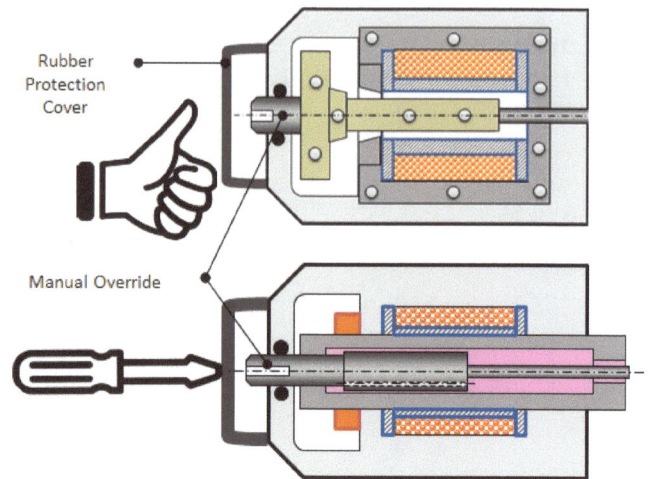

Fig. 7.2 - Manual Override

Note 2 (Solenoid Force vs. Working Temperature): Figure 7.3 shows the force-stroke characteristics of a typical switching solenoid. The figure shows the reduction of the force produced by the switching solenoid due to the increase of the coil temperature.

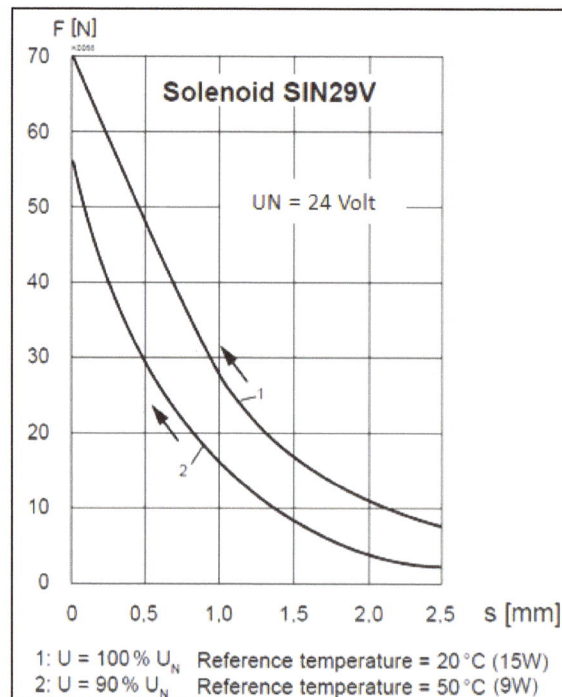

Fig. 7.3 - Typical Force-Stroke Characteristics of a DC Switching Solenoid (Courtesy of Wandfluh)

Note 3 (Cycling rate of Solenoid-Operated Valves):

DC Solenoid Switching Rate: Despite the longer switching time of a DC switching solenoid, it can usually be cycled at a higher rate than an AC switching solenoid. The reason is that a DC switching solenoid does not experience the inrush current phenomenon so that they have better chance to cool down faster during the Off time in the cycle.

AC Solenoid Switching Rate: The high inrush current problem is also a limiting factor in the cycling rate of an AC switching solenoid. Despite the shorter switching time of an AC switching solenoid, it can usually be cycled at a lower rate than a DC switching solenoid. The reason is an AC switching solenoid generates more heat by the inrush current so that it has no chance to cool down between cycles. Consider, for instance, a switching solenoid with a 25% duty cycle as shown in Figure 7.4, the area under each energizing pulse represents heat generation in the switching solenoid. So if the heat can be dissipated, there is no problem. Otherwise, if the heat generated by cycling the switching solenoid exceeds its ability to dissipate the heat, the switching solenoid will overheat and fail. For continuous duty operation, the switching solenoid is capable of dissipating the small amount of heat resulting from the low holding current. The inrush current is the real problem.

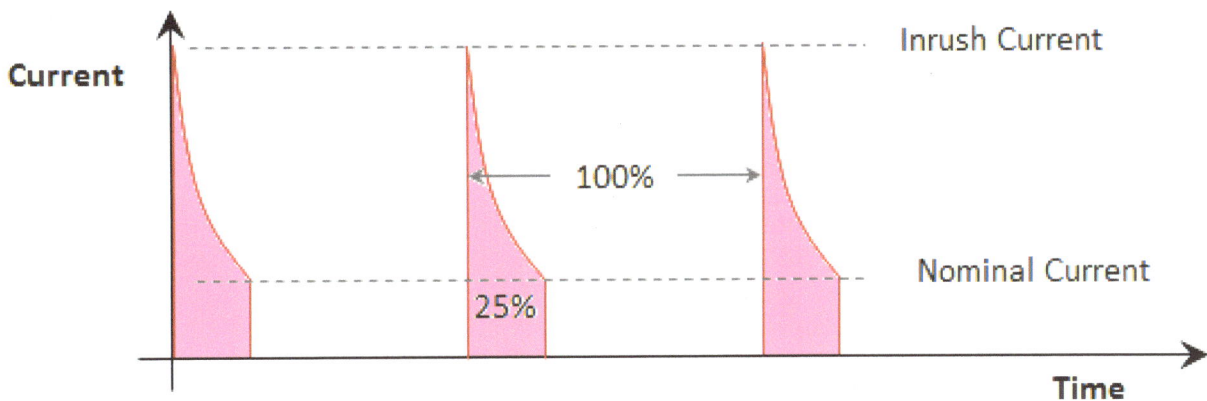

Fig. 7.4 - Switching Rate of an AC Switching Solenoid

Note 4 (EH Valve Ratings):

Nominal Voltages: Switching solenoids are designed for voltages of DC (12V, 24V, and 48V) and AC (110V and 220V). Within limits, DC switching solenoids can also be operated on AC power. Again, this is not recommended by switching solenoid manufacturers. The primary problem here is in the difference in the construction of the two types of switching solenoids. A DC solenoid is not constructed to limit the eddy current and the AC Hum. So, if a DC solenoid operated by AC power, it could be easily burned out.

- Voltage must be at least 90% of the rated current (see Fig. 7.5).
- High voltage heats up the coil.
- Low voltage can't seat the spool completely that resulted in high inrush current.

Nominal Currents: The current absorbed depends on nominal voltage and the size of switching solenoid valves. For EH valves, the current ranges from 1-3 Amps.

Nominal Frequencies for AC Switching Solenoid: In the United States, Canada, and some South American countries, the AC power supplies operate at 60 Hz. In most other countries, 50 Hz is used.

Dual-Frequency Switching Solenoids: Ideally, a switching solenoid should be designed and fabricated for a specific frequency. If a coil is rated for use on a 60 Hz supply is used on 50 Hz, it will draw excess current. If a 50 Hz coil is used on a 60 Hz supply, it will draw less than its rated current, producing a lower electromagnetic force, inrush current period is extended, and the coil becomes less reliable.

Most manufacturers develop dual-frequency switching solenoids that are rated for both 50 and 60 Hz systems. Data sheets for such dual-frequency switching solenoids must reflect that.

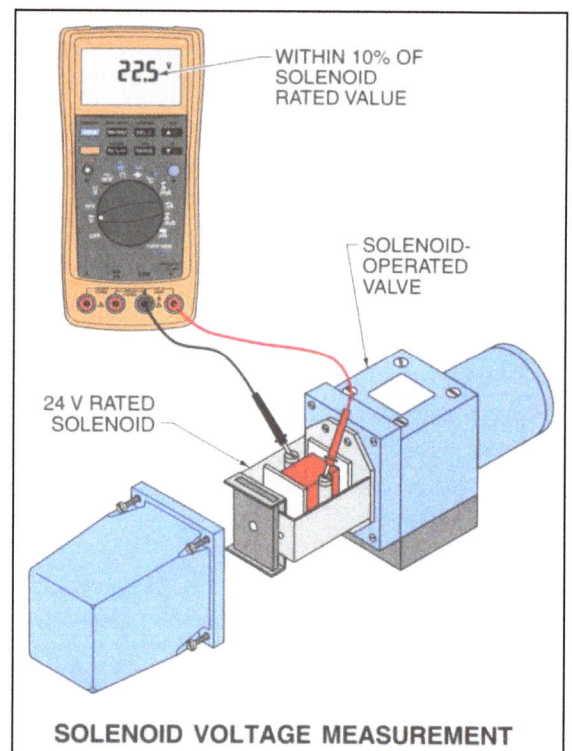

**Fig. 7.5 – Measuring Voltage on Solenoid-Actuated Valve
(Courtesy of American Technical Publishers)**

Note 5 (Solenoid Protection): Careful attention should be given to electrical circuit design to make certain it is not possible for the machine operator, through accident or on purpose, to energize both solenoids of any double solenoid valve at the same time. As shown in Fig. 7.6, mechanical interlocking may be employed to keep one solenoid from being energized while the opposite solenoid is energized.

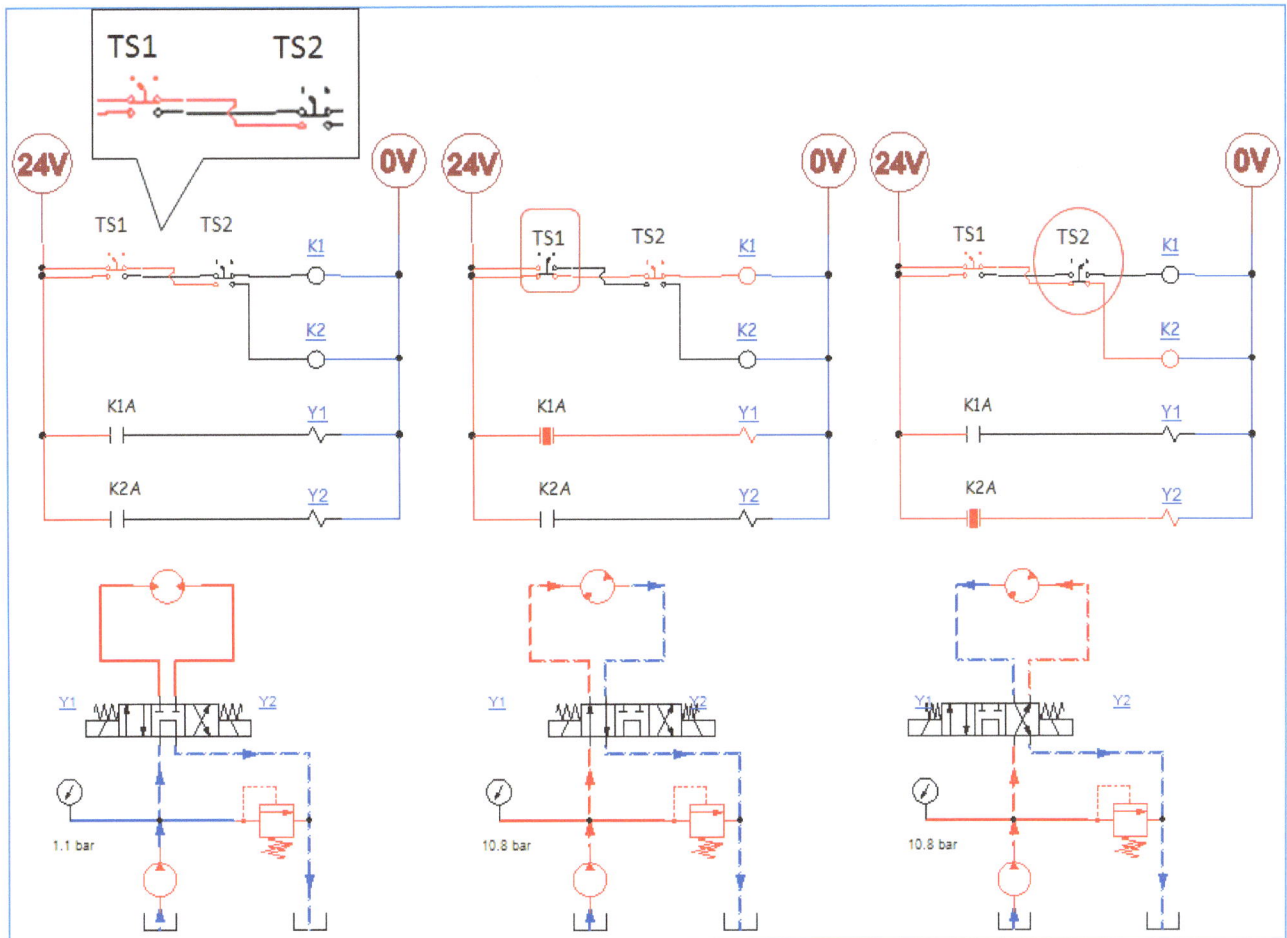

Fig. 7.6 - Mechanical Interlocking for Solenoids Protection

Figure 7.7 shows electrical protection circuit. Even with correct design and interlocking circuits, a relay with sticking contacts or which releases too slowly, could be responsible for a momentary overlap of energization on each cycle. In rush current and consequently heat accumulation over a period of time will eventually cause burn out.

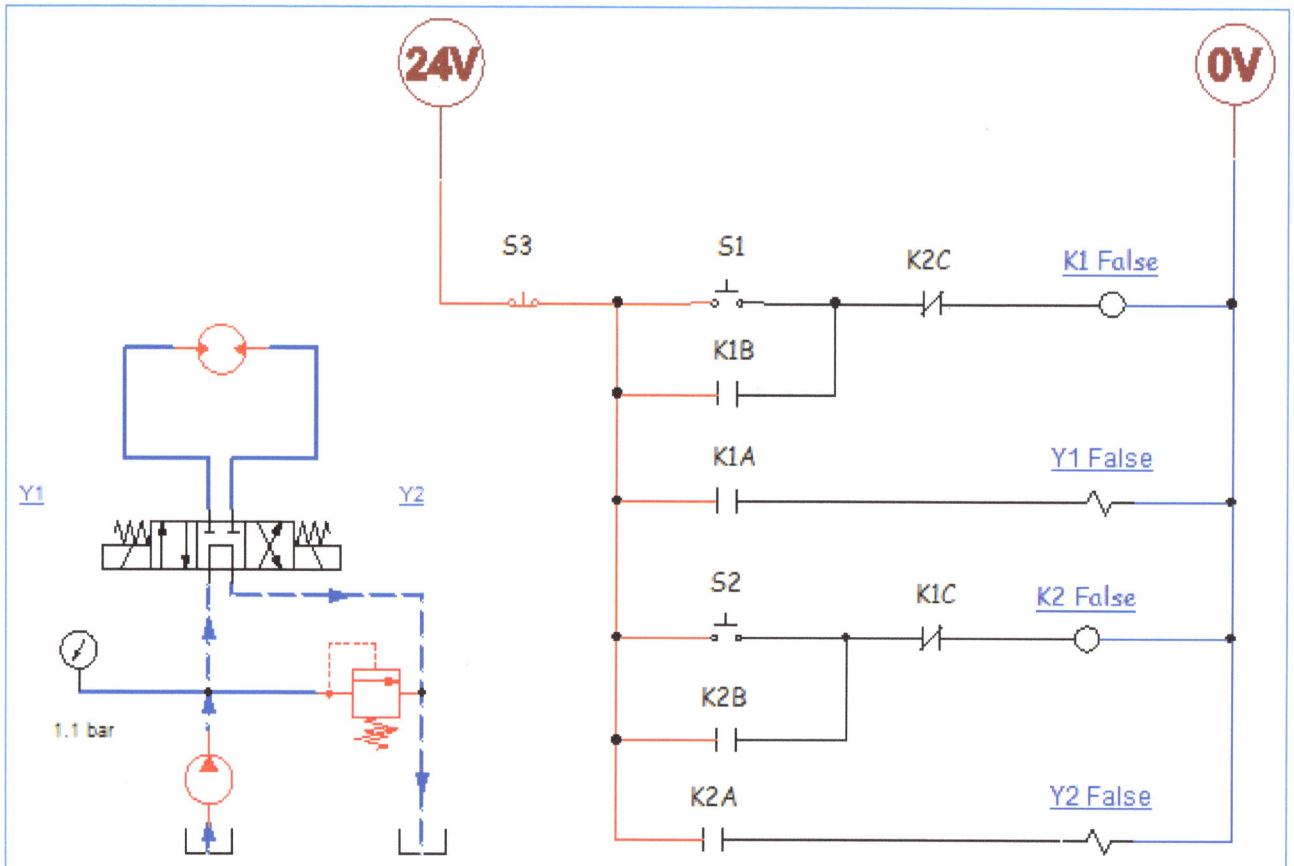

Fig. 7.7 - Electrical Interlocking for Solenoids Protection

Troubleshooting chart shown in Table 7.6 is applied as a preliminary general check on each valve type before getting into the relevant detailed troubleshooting hart.

T-Valve-05: General Valve Troubleshooting	
Valve spool isn't moving.	▪ Hydraulic: ▪ Valve spool is seized due to heavy *contamination*. ▪ Valve *drain* is blocked. ▪ Pilot operated stage isn't working properly. ▪ Mechanical: ▪ Valve body is harshly distorted.
Pilot operated stage doesn't work properly. **(See Note 1).**	▪ Check pilot pressure supply and value. ▪ Check the operation of the pilot valve. ▪ For internal supply of pilot pressure, make sure the central position isn't tandem or open center. Otherwise, add spring loaded check valve. ▪ Make sure pilot stage has floating center. ▪ If main spool has open or tandem center, make sure to use built in check valve. ▪ Pressure fluctuation in tank line or plugged drain line.
Valve leaks externally?	▪ Valve body distorted or cracked. ▪ Static seals failure.
Valve leaks internally?	▪ Valve body distorted. ▪ Worn or broken parts. ▪ Contamination holding valve partially open. ▪ Dynamic seal failures. ▪ Valve is too hot. ▪ Valve spool scored. ▪ Leak at valve seat.
Contaminated Valve?	▪ Disassemble and clean the valve. ▪ If varnish found, investigate sources of varnish formation and act accordingly. ▪ Check the recommended cleanliness level of the valve versus the current cleanliness of the fluid.
Valve body distorted or cracked?	▪ Check if the valve is overtightened to the subplate. ▪ Realign pips connected to the valve to remove mechanical stress.
Weak or broken parts?	▪ Valve subjects to over-pressure or pressure shocks.
Valve seals failure?	▪ Consult Chart: ▪ **"T-Seals-01-Seals Troubleshooting".**
Valve is so hot?	▪ Consult Chart: ▪ **"T-Unit-03-Excessively Hot Unit".**

Table 7.6 – General Valve Troubleshooting Chart

Note 1: Basic Operating of Pilot Operated DCV: As shown in Fig. 7.8, a pilot-operated DCV consists of two stages. The pilot stage is solenoid-actuated, and the main stage is fluidic-actuated. The pilot stage is small while the main stage ports are large enough to handle the large flow. As shown in the figure, control pressure is supplied externally through port **X** or internally through main pressure port **P**. Draining of the control pressure is externally through port **Y** or internally through main tank port **T**. If supply or drain of control pressure have any issue, the whole valve won't properly operate.

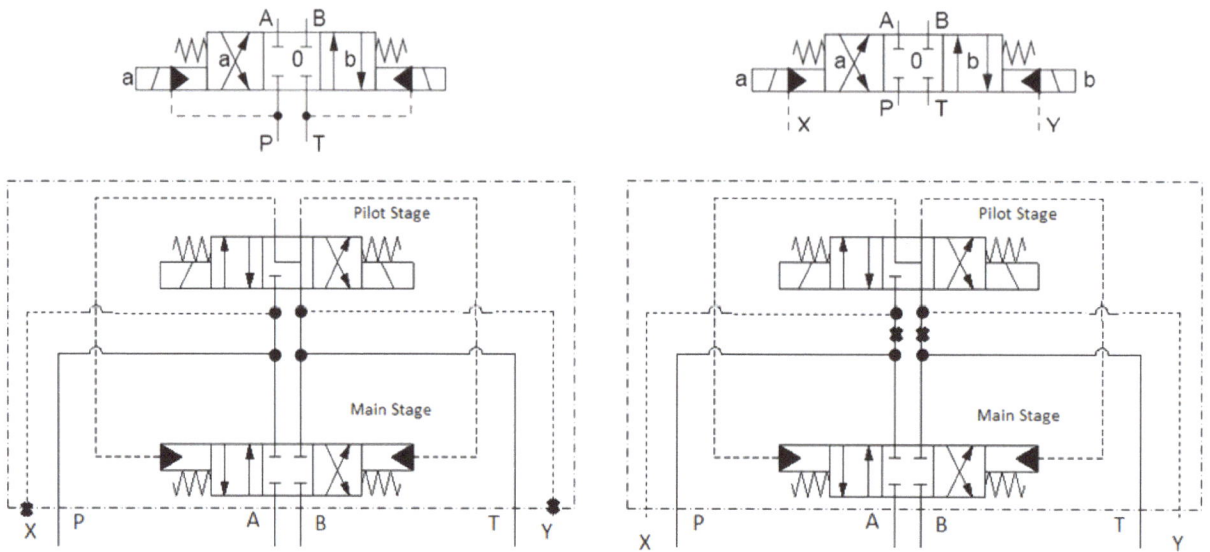

Fig. 7.8 – Pilot-Operated DCV (Courtesy of Bosch Rexroth)

Figure 7.9 shows that a tandem center main spool is used, in which the pressure port is continuously vented through the main spool. In such case, if the control pressure is supplied internally, pump will not be able to drive the load in any case.

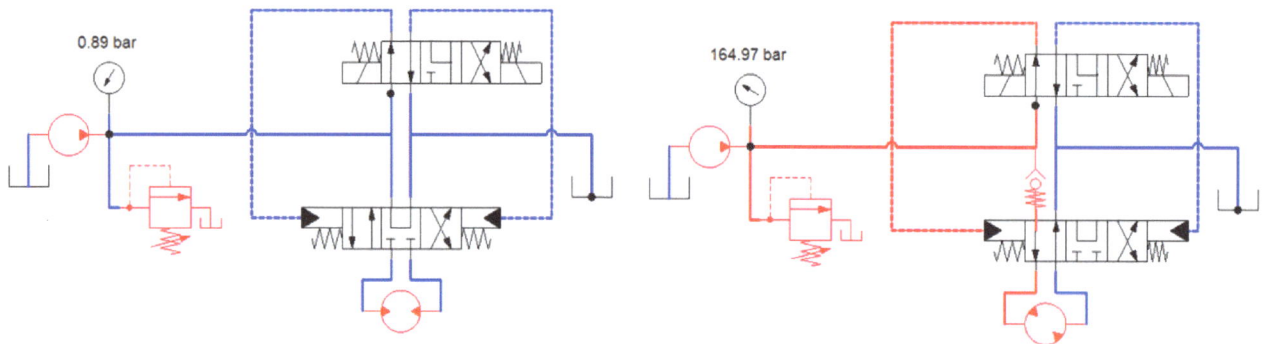

Fig. 7.9 – Use of Spring-Loaded Check Valve with Tandem or Open Center Main Spool

7.3- Hydraulic Valves Failure Analysis

7.3.1- Hydraulic Valves Failure due to Particulate Contamination

Silt Lock: *Silt Lock* (also known as *Contamination Lock*) is an accumulation of silt causing seizure or jamming of components. It is a type of failure that usually doesn't involve wear or permanent internal damage to components, it is rather sudden and unpredictable.

Because of its lack of warning or predictability, silt lock is responsible for a significant number of catastrophic failures in mechanical machinery including even loss of human life. Silt lock has been found to be the root cause of countless failures related to aircraft, spacecraft, passenger cars, elevators, turbine generators, tower cranes, etc.

Silt Lock usually occurs in control valves preventing spool movement from neutral to a shifted position and vice versa. This results in unpredicted actuator movement or failure of the actuator to stop moving.

Electrohydraulic spool valves such as solenoid, pulse-width modulated (PWM), proportional control, and servo valves are sensitive to silt lock. As shown in Fig. 7.10, silt particles can enter the clearances between the spool and bore in the leakage path. This increases the static friction of the spool when the valve is actuated. This reduces the valve dynamic response and causes a *stick-slip* movement, which is also known as a *hard-over* condition.

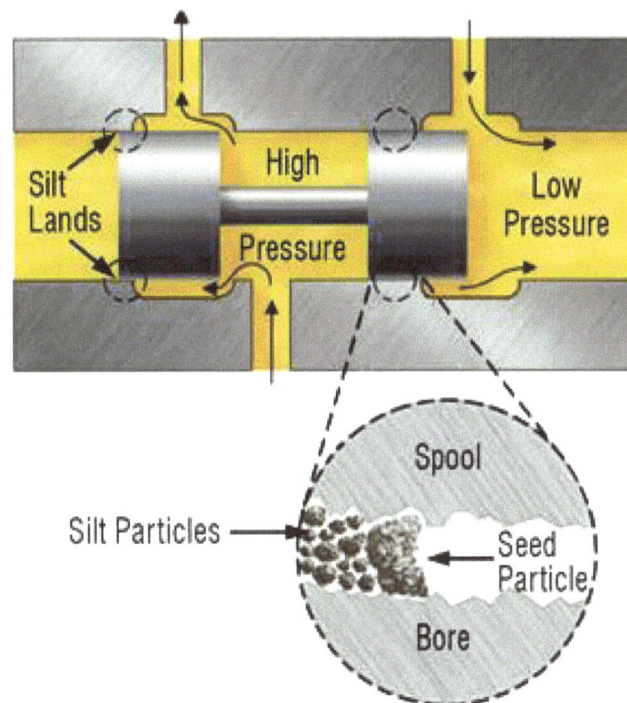

Fig. 7.10- Silt Lock in Spool Valves (Courtesy of Noria Corporation)

Worn Spool Valves due to Abrasive Contaminants: Figure 7.11 shows how a spool is affected by abrasive wear. Spools in such conditions are not reusable.

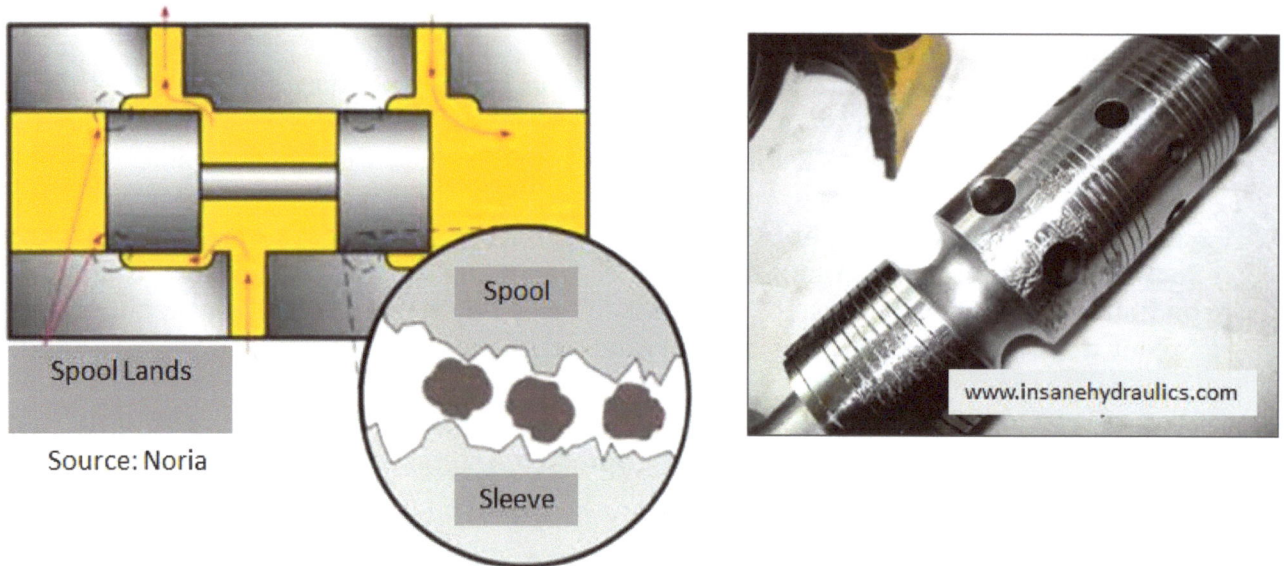

Fig. 7.11- Commonly Worn Surfaces in Spool Valves

Worn Poppet Valves due to Abrasive Contaminants: Figure 7.12 shows how a poppet is affected by abrasive wear causing the valve seat improperly resulting in leakage.

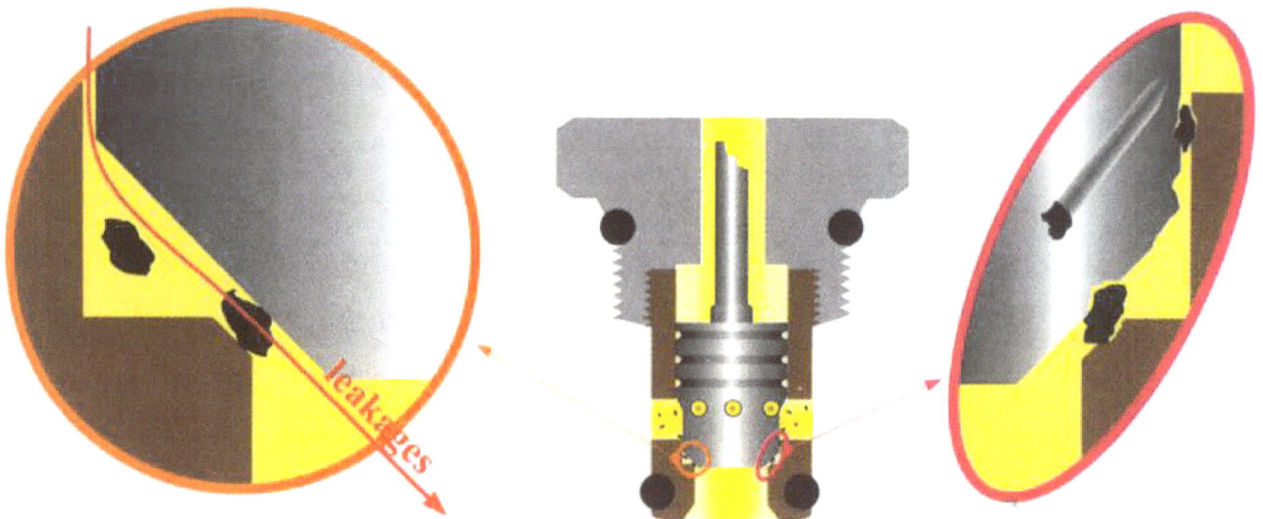

Fig. 7.12- Commonly Worn Surfaces in Poppet Valves (Courtesy of ASSOFLUID)

Servo Valve Nozzle Blockage and Wear: As shown in Fig. 7.13, a jet-pipe servo valve fails passively or safely as compared to the flapper nozzle. To explain that, assume a piece of contamination blocked the jet pipe, none of the receiving holes will receive any power. Therefore, the main spool stays in neutral position and the load as well. On the other hand, if one of the nozzles is blocked by contamination, this is equivalent to a situation where the torque motor is energized with maximum signal. In such a case the spool will suddenly shift to full stroke and the load will suddenly receive maximum hydraulic power.

The figure also shows that erosion in a jet pipe stage happens almost in a symmetric way between the two receiving holes. That distributes the deviation in the valve characteristics evenly. In a flapper-nozzle, if the erosion in one side of the flapper is different from the other side by 30 microns, this will result in a 50% change in the valve flow gain and doubles the valve leakage.

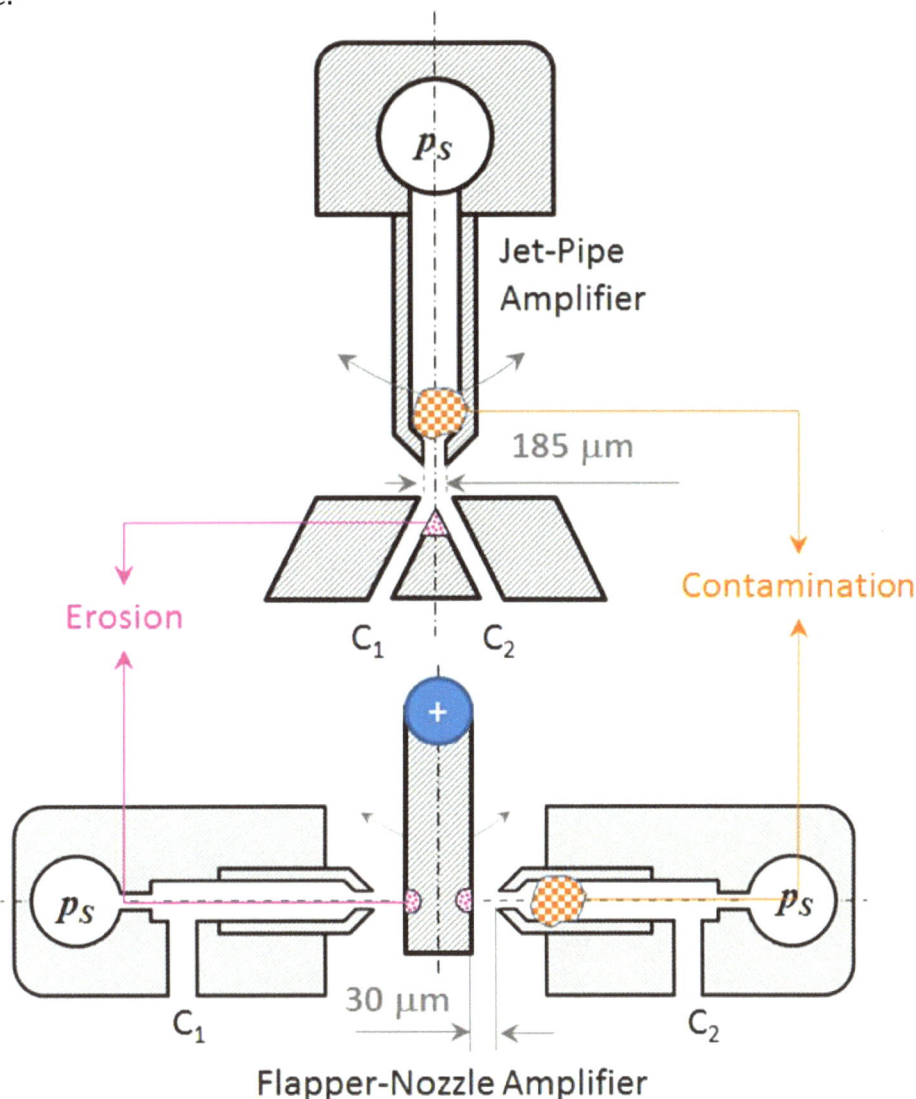

Fig. 7.13- Jet-Pipe versus Flapper-Nozzle Pilot Stage

7.3.2- Hydraulic Valves Failure due to Chemical Contamination

Figure 7.14 shows the process of *Varnish* formation. Varnish is a thin, insoluble, non-wipeable, gummy, and sticky film deposit on metal surfaces. The figure shows examples of varnish formation on various machine components within a hydraulic system.

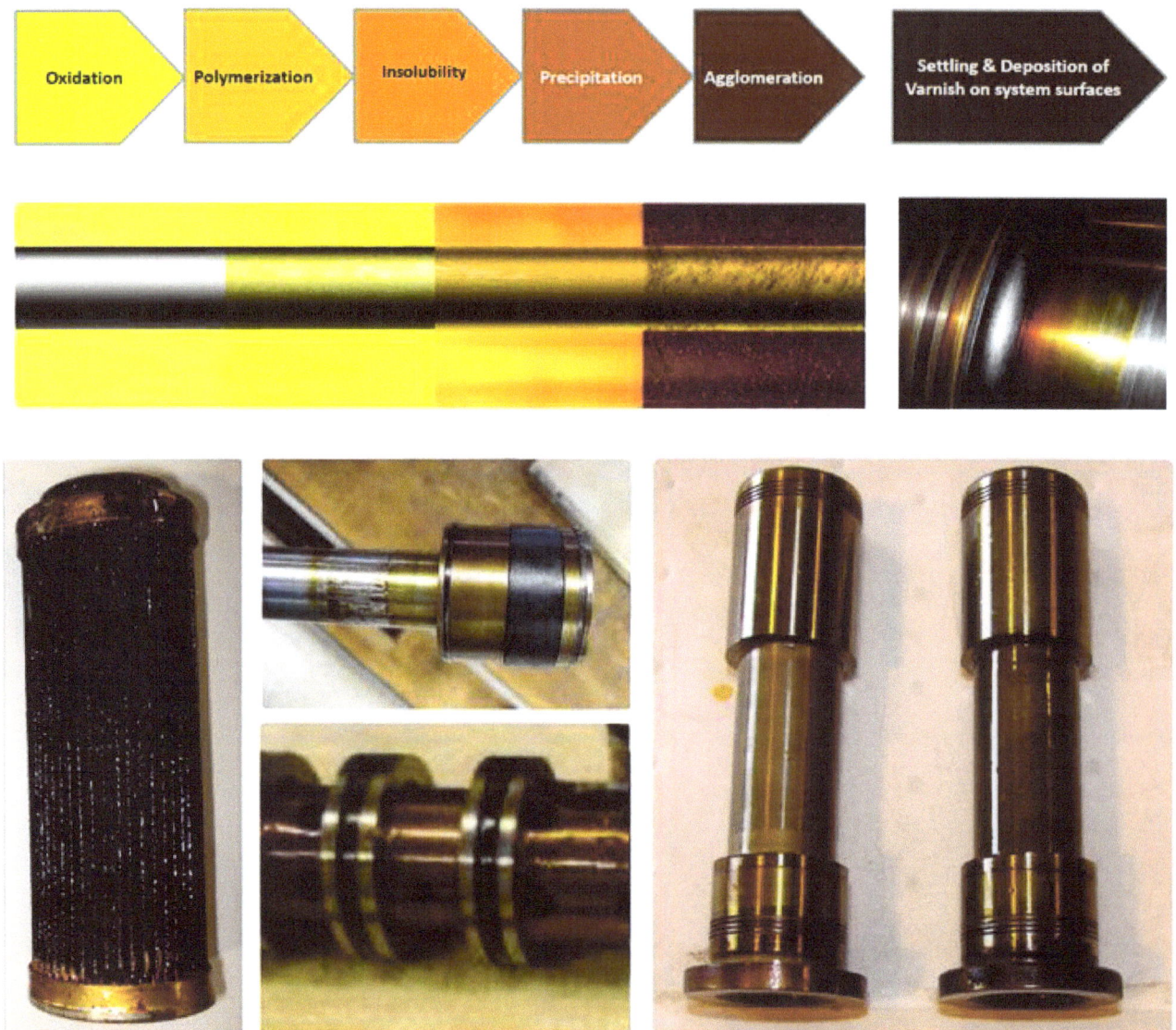

Fig. 7.14- Varnish Formation within Hydraulic Systems (Courtesy of C.C. Jensen Inc.)

As shown in Fig. 7.15, varnish creates a sticky layer. This layer attracts the abrasive particles of all sizes creating a sandpaper grinding surface which radically speeds up machine wear. As shown in Fig. 7.16, varnish can easily block fine tolerances, making spool valves (e.g. directional control valves) seize.

Fig. 7.15- Varnish Sticky Layer Attracts Abrasive Particles (Courtesy of C.C. Jensen Inc.)

Fig. 7.16- Varnish Sticky Layer Seizes Valve Spools (Courtesy of C.C. Jensen Inc.)

7.3.3- Solenoid Burn Out

Causes of Solenoid Burnout are voltage too low or too high, high ambient temperature, cycling too fast, Inrush Current due to valve spool is blocked.

Consequences of the *Inrush Current:*
Solenoid valve coil burnout is a more common problem in industrial hydraulics than in mobile hydraulics. The reason is while DC solenoids are rarely burnout, AC solenoids are mainly used for industrial hydraulic. In DC solenoids, the electrical current is constant for a given voltage. Because the resistance of the coil is constant. When voltage is applied to a DC solenoid, the current draw rises from zero to the maximum value which can pass through the coil-regardless of the position of the solenoid's armature in relation to its coil. AC coils behave very differently to DC coils.

The resistance or *impedance* of an AC coil is lowest when the solenoid is open, i.e. when the armature is out. Impedance increases as the armature is pulled into the closed position. As a result, the current draw of an AC solenoid is highest when the solenoid is open (armature out) and lowest when the solenoid is closed (armature in). The high current draw of an open AC solenoid is known as *inrush* current. And the current draw when the solenoid is closed is called *holding* current. AC solenoids can only dissipate the heat generated by their holding current. This means it's very important for the armature *to* close completely when an AC solenoid is energized.

In some cases, the armature may not be able to move fully into the coil and close the gap due to some external reasons such as mechanical failure, contamination seizure, and high flow forces in undersized valves. In such cases, during the early stage of the armature movement where the magnetic field is still weak, the current will continue to rush into the coil. The temperature of the coil will then rise rapidly until it exceeds the rated temperature of the winding insulation. This will eventually create a short circuit across the turns of the coil winding leading to a catastrophic failure of the coil. As shown in Fig 7.17, the resulting heat could melt the insulation and burn the coil in minutes. It also boils the oil around the core in a wet type switching solenoid.

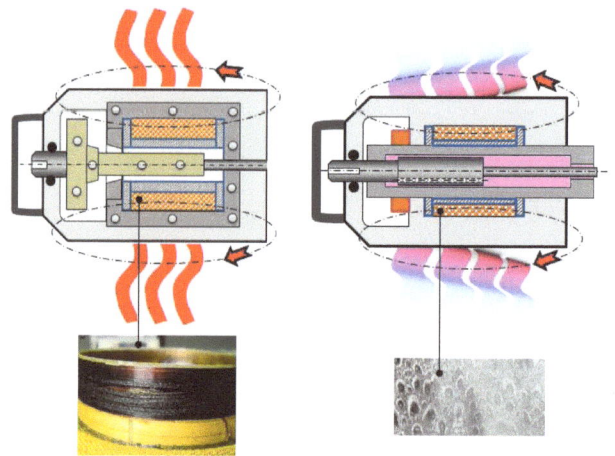

Fig. 7.17 - Coil Overheating Due to Inrush Current

7.3.4- Spool Failure due to Dry Operation

Figure 7.18 shows that energizing a switching or a proportional valve without oil supply to the main pressure port results in high spool-sleeve friction. In servo valves, if the torque motor is tested without fluid flow through the flapper nozzle stage, feedback spring is severely stretched.

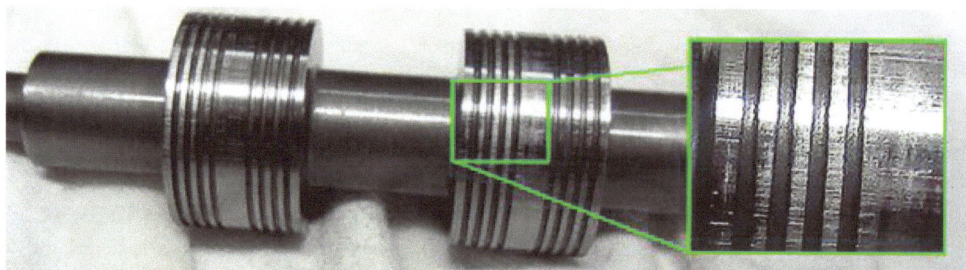

Fig. 7.18- Failed Spool due to Dry Operation (Courtesy of H & P Magazine.)

Chapter 8

Troubleshooting and Failure Analysis of Accumulators

Objectives

This chapter discusses hydraulic *accumulators* inspection, troubleshooting, and failure analysis. In this chapter, a troubleshooting chart for accumulator faults is presented. The chapter also presents examples of defective accumulators caused by various reasons such as vessel explosion due to material defects and pressure shocks etc.

Brief Contents

8.1- Hydraulic Accumulators Inspection

8.2- Hydraulic Accumulators Troubleshooting

8.3- Hydraulic Accumulators Failure Analysis

Chapter 8: Troubleshooting and Failure Analysis of Accumulators

8.1- Hydraulic Accumulators Inspection

A hydraulic *accumulator* is a capacitive device that can store energy. Therefore, for safety, it is mandatory to review manufacturer's instructions in any case that the accumulator is inspected for faults or repaired. Accumulators could be of a piston-type, bladder-type, or diaphragm-type. Volume 5 of these series of textbooks provides information about the construction and operating principles of accumulators. Table 8.1 shows typical inspection sheet for a hydraulic accumulator.

Hydraulic Accumulator Inspection Sheet	
Manufacturer	
Model #	
Serial #	
Location	
Accumulator Type	▪ ☐ Piston ☐ Bladder ☐ Diaphragm
Accumulator Design Values	▪ Nominal Volume: ( ) ▪ Pre-charge Pressure: ( ) ▪ Min System Pressure: ( ) ▪ Max System Pressure: ( )
Current Charge Pressure	▪ Pre-charge Pressure: ( )
Conditions of Bladder/Diaphragm	
Conditions of Accumulator Body	
State of Gas Charge Valve	
Other Notes:	

Table 8.1 – Hydraulic Accumulators Inspection Sheet

8.2- Hydraulic Accumulators Troubleshooting

Table 8.2 shows troubleshooting guidelines for accumulators.

T-Accumulator-01: Accumulator Troubleshooting	
Response of accumulator is slow?	• Pre-charge gas pressure improperly adjusted? • Piston binding or weak spring. • Relief valve set too low.
Accumulator fails to absorb shocks?	• Pre-charge gas pressure is lost or too high. • Diaphragm or bladder ruptured. • Piston seized.
Accumulator isn't properly charged with oil.	• Check maximum accumulator's pressure setting. • Check pre-charge gas pressure. • Check leaking accumulator oil discharge valve. **(See Note 1).**
Pre-charge gas pressure improperly adjusted?	• Check for possible gas leak from the accumulator. • Check if the bladder or diaphragm is ruptured. • Check the conditions of seals in piston accumulators. **(See Note 2).**
Maximum accumulator's pressure improperly adjusted?	• Review maximum pressure and adjust PRV accordingly.
Leaking accumulator oil discharge valve?	• Check the if the valve is tightened properly.
Accumulator charging time is unusually long.	• Pump undersized. • Pre-charge gas pressure too low. • PRV is set too low or partially stuck open. • Leaking oil-discharge valve or partially opened.
Loose body assembly bolt?	• Torque bolts as per the manufacturer's instructions.
Piston binding or seizure in a piston accumulator?	• Discharge the gas and oil. • Disassemble and repair.
Bladder/Diaphragm is ruptured?	• Check if the compression ratio exceeds the design value **(See Note 3).**

Table 8.2– Accumulator Troubleshooting Chart

Note 1 – Accumulator Mounting Manifold: As shown in Fig. 8.1, for safety operation, accumulators must be mounted on mounting manifold that contains the following elements:
- Isolation valve to connect/disconnect the accumulator with the system.
- Manual discharge valve to discharge oil stored in the accumulator to the tank if needed.
- Solenoid-Actuated oil discharge valve (optional) for automatic oil drain in case of turning off the machine or emergency cases such as power outage.
- Maximum gas pressure adjustment pressure relief valve.

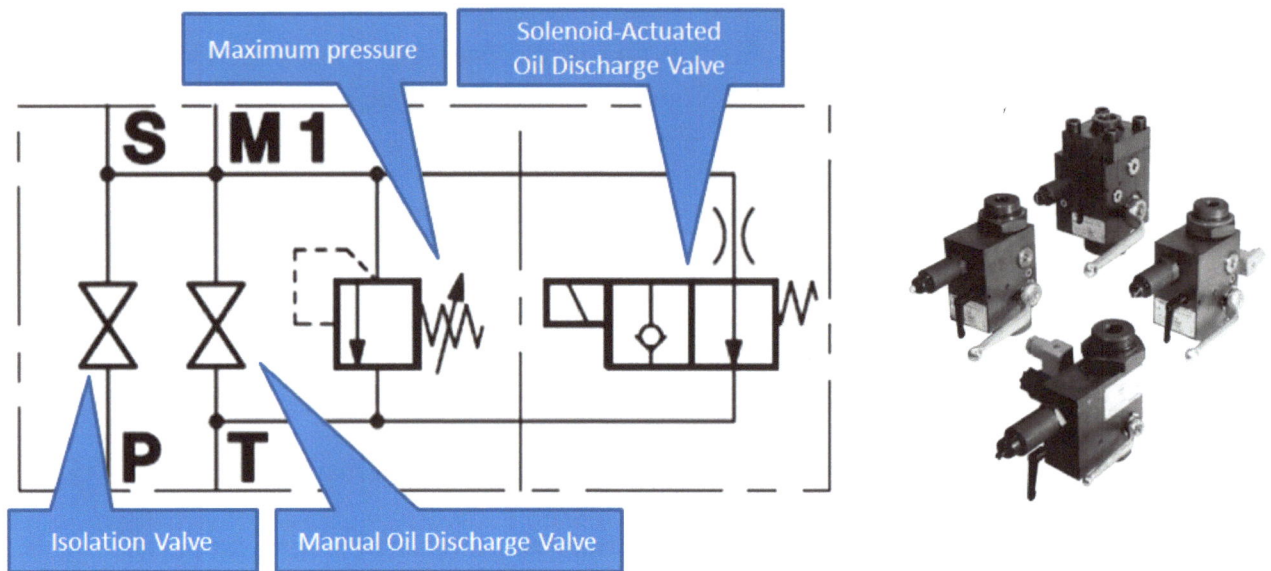

Fig. 8.1 – Mounting Manifold for Accumulators

Note 2 – Accumulator Mounting Precharge:

Pre-charge pressure too high: This may cause operational problems or damage to accumulators. With a piston accumulator the piston will travel too close to the hydraulic end cap and the piston could bottom out, reducing output and eventually damaging the piston and piston seal. The piston can often be heard bottoming, warning of impending problems. In a bladder accumulator the bladder can be driven into the poppet assembly when discharging. This could cause a fatigue failure of the poppet spring assembly, or even a pinched bladder. Excessive pre-charge pressure is the most common cause of bladder failure.

Pre-charge pressure is too low: This can also cause operating problems and subsequent accumulator damage. With no pre-charge in a piston accumulator, the piston will be driven into the gas end cap and will often remain there. Usually, a single contact will not cause any damage, but repeated impacts will eventually damage the piston and seal. Conversely, for a bladder accumulator, too low or no pre-charge can have rapid and severe consequences. The bladder will be crushed into the top of the shell and can extrude into the gas stem and be punctured. One such cycle is sufficient to destroy a bladder. Overall, piston accumulators are generally more tolerant of incorrect pre-charging pressures.

Note 3 – Accumulator Compression Ratio: As shown in Fig. 8.2, before assembling the accumulator in the system, it will be charged with nitrogen under pressure p_1 that is specified by the end user. The Nitrogen occupies the full nominal volume of the accumulator V_1 under the initial gas pressure. When the accumulator is used in the system, a minimum pressure p_2 must be slightly higher than the initial gas pressure in order to lift the separator a bit from the fluid port. If the system pressure increases above p_2, the fluid is forced into the accumulator. As a result, the gas is compressed, and its pressure increases. When the fluid volume in the accumulator increases, gas pressure increases up to a maximum pressure p_3 that is determined by the configuration of the accumulator and the operating system.

If the fluid pressure is reduced below gas pressure, the gas expands, and the fluid is discharged out of the accumulator. The maximum oil volume, ΔV_{max} is stored in the accumulator at pressure between p_2 and p_3. The ratio $p_3 : p_2$, is defined as the *Compression Ratio*.

Piston accumulators can afford compression ratio up to 9. Bladder accumulators can afford compression ratio up to 4-5. Diaphragm type accumulator can afford compression ratio up to 2. Exceeding compression may damage the separator, particularly bladders and diaphragms.

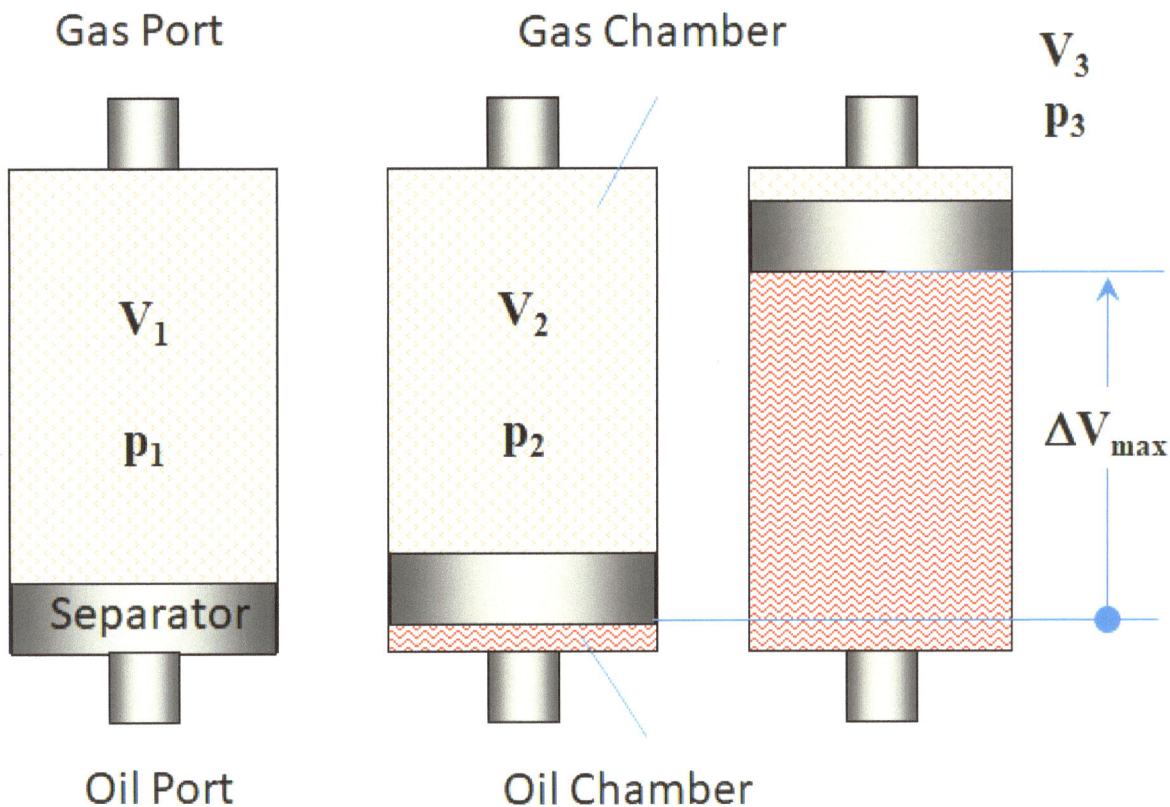

Fig. 8.2 - Operating Principle of Accumulators

8.3- Hydraulic Accumulators Failure Analysis

The following incidents presents typical failure cases of accumulators.

Accumulator Failure due to Material Defects: The failure shown in Fig. 8.3 was attributed to a material defect in the body wall of the accumulator.

Accumulator Failure due to Exceeding Compression Ratio: As shown in Fig. 8.4, the accumulator is failed due to pressure shock that exceed the maximum pressure.

Accumulator Failure due to Pinched Bladder: As shown in Fig. 8.5, in a bladder without cushioning cup, the bladder wraps around the poppet while fluid is flowing out. Then the poppet closes on the bladder pinching it. Cushioning cup helps avoid such a failure.

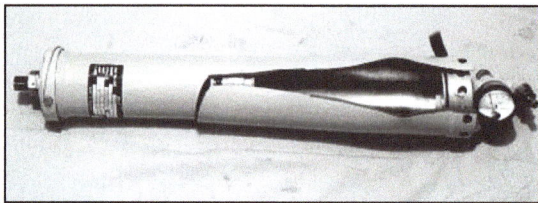

Fig. 8.3 – Accumulator Failure due to Material Defect

Fig. 8.4 – Accumulator Failure due to Exceeding the Compression Ratio

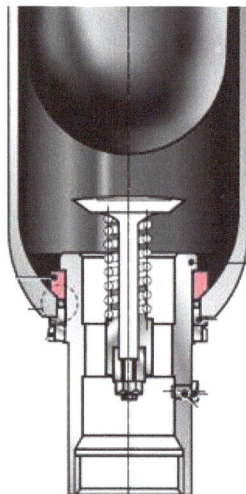

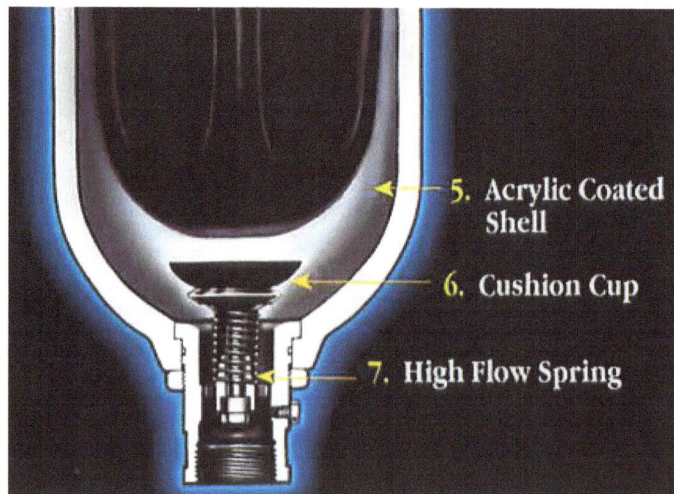

5. Acrylic Coated Shell

6. Cushion Cup

7. High Flow Spring

Fig. 8.5 – Accumulator Failure due to Accumulator Rupturing

Chapter 9

Troubleshooting and Failure Analysis of Reservoirs

Objectives

This chapter discusses hydraulic *reservoirs* inspection, troubleshooting, and failure analysis. In this chapter, a troubleshooting chart for reservoir faults is presented. The chapter also presents examples of defective reservoirs.

The following topics are discussed in Chapter 2 in Volume 4 "Hydraulic Fluids Conditioning" of this series of textbooks:
- Contribution of Hydraulic Reservoirs.
- Configurations of Hydraulic Reservoirs.
- Construction of Hydraulic Reservoirs.
- Design of Hydraulic Reservoirs.
- Hydraulic Reservoir Design Case Study.

The following topics are discussed in Chapter 9 in Volume 5 "Maintenance and Safety" of this series of textbooks:
- BP-Reservoirs-01-Selection and Replacement.
- BP-Reservoirs-02-Maintenance Scheduling.
- BP-Reservoirs-03-Installation and Maintenance.

Brief Contents

9.1- Hydraulic Reservoirs Inspection

9.2- Hydraulic Reservoirs Troubleshooting

9.3- Hydraulic Reservoirs Failure Analysis

Chapter 9: Troubleshooting and Failure Analysis of Reservoirs

9.1- Hydraulic Reservoirs Inspection

A hydraulic *reservoir* is an essential component that has several duties beyond just hosting the required volume of oil. Hydraulic reservoirs are designed to help in oil cooling, cleaning and deaeration. Volume 4 of these series of textbooks provides guidelines for reservoir design.

Table 9.1 shows typical inspection sheet for hydraulic reservoirs.

Hydraulic Reservoirs Inspection Sheet	
Manufacturer	
Model #	
Serial #	
Location	
Reservoir Type	▪ ☐ L-Shaped ☐Foot-Mounted ☐Overhead ☐Compact ▪ ☐ Open to Atmosphere ☐Pressurized
Reservoir Size	▪ Liters:( ) Gallons:( )
Current Charge Pressure	▪ Pre-charge Pressure: ( )
Conditions of Breather Filter	
Conditions of Connected Lines	
Conditions of Gaskets	
Conditions of Inside	
Other Notes	

Table 9.1 – Hydraulic Reservoirs Inspection Sheet

9.2- Hydraulic Reservoirs Troubleshooting

Table 9.2 shows troubleshooting guidelines for accumulators.

T-Reservoirs-01: Reservoir Troubleshooting	
Leaking reservoir?	▪ Check gaskets, flanges, other line connections.
Filter breather clogged?	▪ Replace clogged filter.
Hydraulic fluid aerated?	▪ Review reservoir design **(See Note 1)**. ▪ Consult Chart: ▪ **"T-System-01-Fluid Aeration"**.
Hydraulic fluid contaminated by water?	▪ Check the condition of water absorption cartridge on breather filter. ▪ Check the condition of oil-water heat exchanger. ▪ Cover the reservoir from rain. ▪ Apply appropriate water separation method.

Table 9.2– Reservoir Troubleshooting Chart

Note 1: Reservoir Design Tips to Help Fluid Deaeration: As shown in Fig. 9.1, placing a baffle plate and a screen between the return oil and the suction strainer help removing air. Detailed information is found on Volume 4 of this series of textbooks.

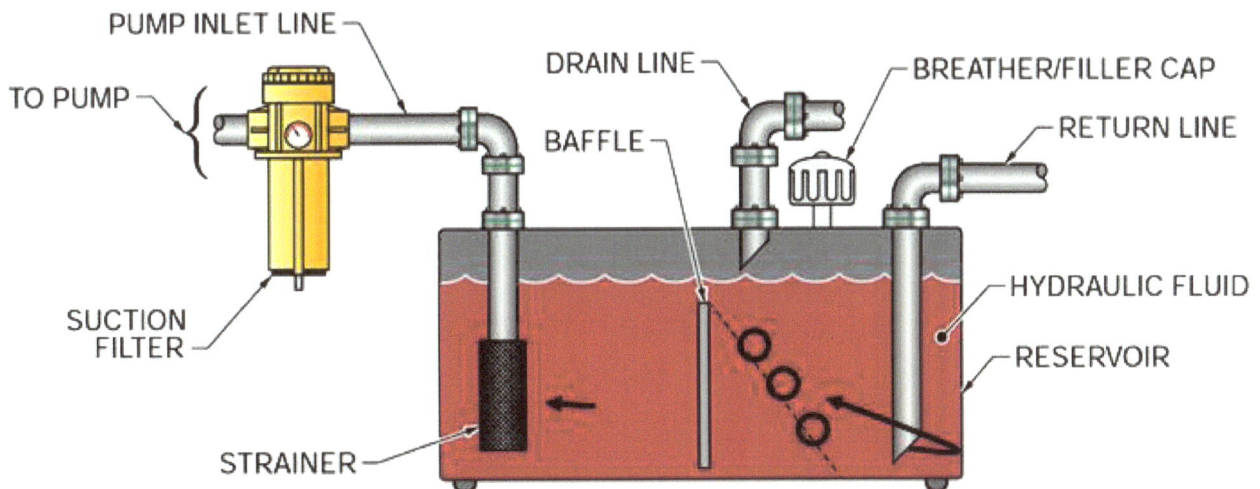

**Fig. 9.1- Role of Hydraulic Reservoir in Fluid Deaeration
(Courtesy of American Technical Publishers)**

9.3- Hydraulic Reservoirs Failure Analysis

Leaking Gasket: As shown in Fig. 9.2, in a hydraulic reservoir of a tractor machine, the access hole to the hydraulic reservoir is slowly leaking/dripping. The tank has no damage, but brand-new gasket has been replaced twice, where none of them were dry or damaged. It was eventually found improper gasket material was used that is incompatible with the hydraulic fluid.

www.tractorbynet.com

Fig. 9.2- Leaking Gasket of Hydraulic Reservoir

Heavily Contaminated Reservoir: As shown in Fig. 9.3, a hydraulic reservoir is heavily contaminated. Rust, stains, and varnish are accumulated on the side walls. The overall system required thorough flushing after reservoir cleaning and sandblasting.

Fig. 9.3- Heavily Contaminated Reservoir

Damaged Attachments and Filling Cap: As shown in Fig. 9.4, the reservoir is carelessly treated. As shown in the figure, some attachments are bent, and filling cap thread is damaged.

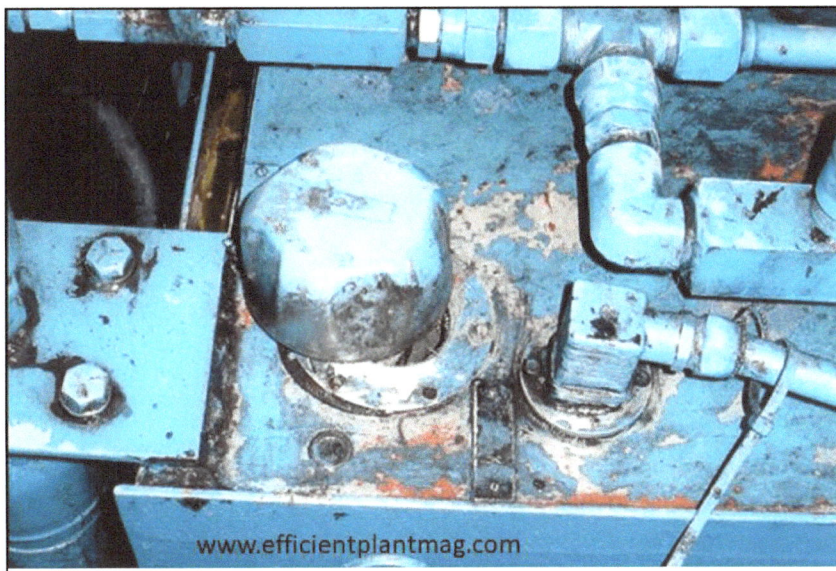

Fig. 9.4- Damaged Attachments and Filling Cap

Chapter 10

Troubleshooting and Failure Analysis of Transmission Lines

Objectives

This chapter discusses hydraulic *transmission lines* inspection, troubleshooting, and failure analysis. In this chapter, a troubleshooting chart for transmission line faults is presented. The chapter also presents examples of defective transmission lines.

The following topics are discussed in Chapter 3 in Volume 4 "Hydraulic Fluids Conditioning" of this series of textbooks:
- Basic Types and Contribution of Hydraulic Transmission Lines.
- Sizing of Hydraulic Transmission Lines.
- Rated Pressures for Hydraulic Lines.
- Hydraulic Pipes.
- Hydraulic Tubes.
- 3.6- Hydraulic Hoses.
- Flanges for Transmission Line Connections.
- Rubber Expansion Fittings.
- Test Points.
- Pressure Measurement Hoses.
- Manifolds.

The following topics are discussed in Chapter 10 in Volume 5 "Maintenance and Safety" of this series of textbooks:
- 10.1-BP-Transmission Lines-01-Selection and Replacement.
- 10.2-BP-Transmission Lines-02-Maintenance Scheduling.
- 10.3-BP-Transmission Lines-03-Installation and Maintenance.
- 10.4-BP-Transmission Lines-04-Standard Tests and Calibration.
- 10.5-BP-Transmission Lines-05-Transportation and Storage.

Brief Contents

10.1- Hydraulic Transmission Lines Inspection
10.2- Hydraulic Transmission Lines Troubleshooting
10.3- Hydraulic Transmission Lines Failure Analysis

Chapter 10: Troubleshooting and Failure Analysis of Transmission Lines

10.1- Hydraulic Transmission Lines Inspection

Hydraulic *transmission lines* could be pipes, tubes or flexibles hoses. They are used to transmit fluid power between hydraulic components. Failure of any transmission line causes consequences that range from minor problem to loss of life. Volumes 4 and 5 of this series of textbooks provides guidelines for transmission lines design, maintenance and safety.

Table 10.1 shows typical inspection sheet for transmission lines.

Hydraulic Transmission Lines Inspection Sheet	
Manufacturer	
Model #	
Serial #	
Location	
Transmission Line Type	☐ Pipe ☐Tube ☐Hoses
Transmission Line ID	mm:( ) inches:( )
Transmission Line Length	
Transmission Line Ends	
Conditions of Transmission line	
Conditions of Transmission line	▪ Hose kinks and twists. ▪ Hose cracks from minimum bend radius exceeded. ▪ Hose brittleness or loss of flexibility. ▪ Hose frayed protective layers. ▪ Hose broken reinforcement layers. ▪ Hose outer cover pulled back from coupling ends. ▪ Hose rusted, broken, or loosen ends joints. ▪ Hose abrasion.
Other Notes	

Table 10.1 – Hydraulic Transmission Lines Inspection Sheet

10.2- Hydraulic Transmission Lines Troubleshooting

Problem of leaking transmission line is beyond just loss of fluid. The real cost of leaking fluid is the cost of:

- Make-up the fluid.
- Cleaning-up the mess.
- Proper disposal of the spilled fluid in accordance with the local state and federal regulations.
- Contamination ingression.
- Possible safety liability.

To get a better sense of the real cost of leakage, a study has been done to calculate the cost of a fitting that is leaking six drops of oil per minute. The cost was found to be nearly $1000 per year.

Table 10.2 shows troubleshooting guidelines for transmission lines.

T-Transmission Lines-01: Transmission Lines Troubleshooting	
Leaking transmission line? **(See Note 1)**	▪ Line Body: ▪ Check if any part of the line body is damaged. ▪ Check if the line is subjected to mechanical stresses. ▪ Check if the line is over pressurized. ▪ Check if the line wasn't cleaned before assembly **(See Note 2).** ▪ Check minimum bend radius for hoses. ▪ Check if a hose is twisted or stretched. ▪ Check if a hose was subjected to abrasion. ▪ Line Fittings: ▪ Check if line ends are damaged. ▪ Check if line ends are standard and of good quality. ▪ Check tightening torque of line ends. ▪ Flare has cracks or embedded dirt. ▪ Tube is not properly aligned with fitting. ▪ There is no sealant used. ▪ Fitting threads are distorted. ▪ O-Ring leak. ▪ Hydraulic Fluid: ▪ Verify that the fluid is compatible with the inner tube, the outer cover, fittings, and O-rings.

Table 10.2– Transmission Lines Troubleshooting Chart

Note 1: Transmission Line Leakage Inspection: Figure 10.1 shows an effective method of leakage inspection by blending special dye with the fluid. By combining a small amount of concentrated dye to a fluid system with a specific wavelength lamp, even the smallest leaks become easily visible. Fluid leakage test kit was previously shown in chapter 2 (Fig. 2.8).

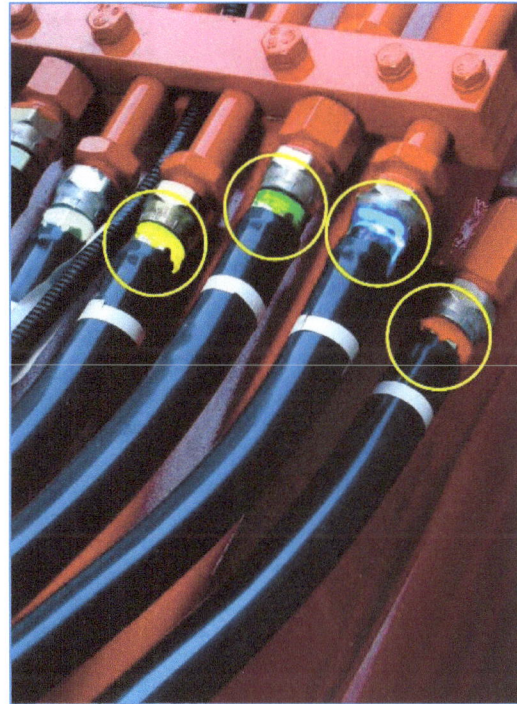

Fig. 10.1– Effective Method for Transmission Line Leakage Inspection (Courtesy of Spectroline)

Note 2: Hose Leakage due to Line Not Cleaned Before Assembly: Contamination can cause several problems for a hydraulic hose assembly. As shown in Fig. 10.2, When cutting a hose, metal particles and debris can settle inside the hose if not properly flushed. This abrasive debris causes small fractures to develop between fitting and hose assembly, resulting in leakage. To prevent hose failures from contamination, the hose must be properly cleaned before inserting the fittings. After the fittings are crimped, be sure to cap the ends in order to keep the hose clean and avoid recontamination during transportation.

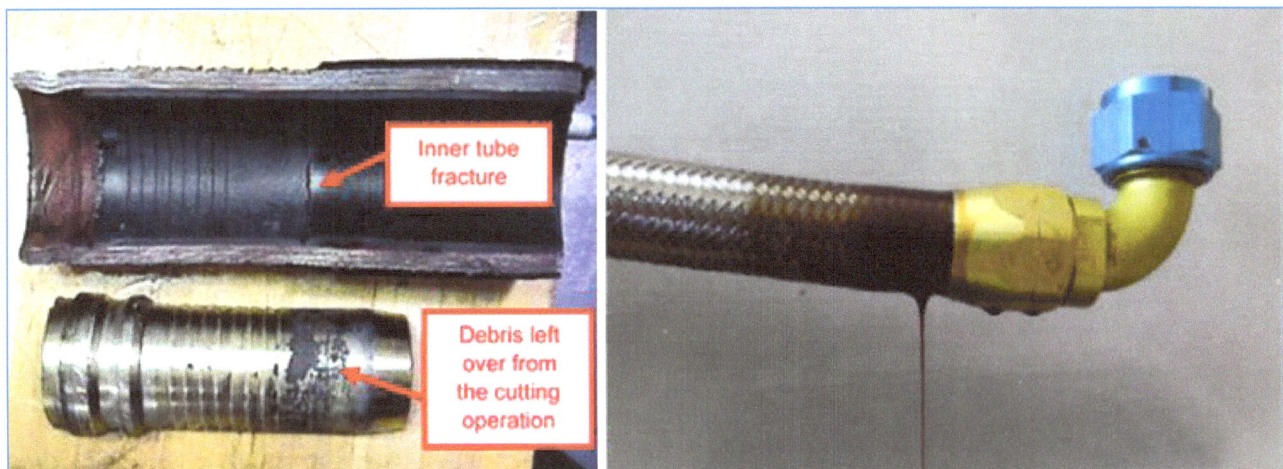

Fig. 10.2– Hose Leakage due to Line Not Cleaned Before Assembly (Courtesy of Parker)

10.3- Hydraulic Transmission Lines Failure Analysis

Hose Cover Abrasion: Figure 10.3 shows hose cover abrasion due to rubbing with a metallic sharp edge. The figure shows best practices of resolving this problem by mounting the hose away from moving elements.

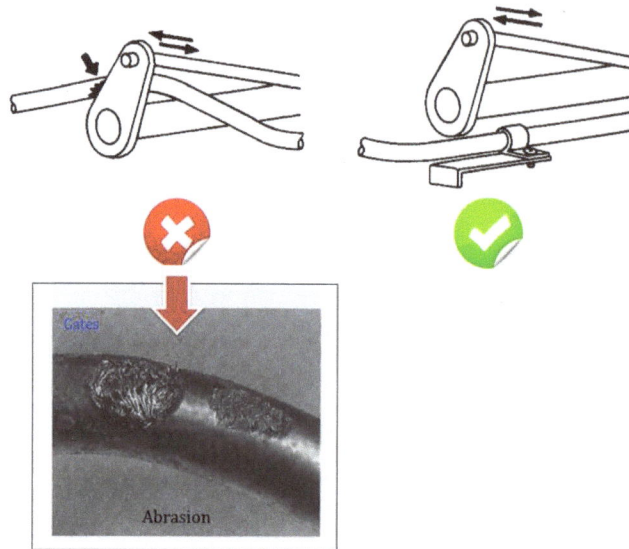

Fig. 10.3– Hose Cover Abrasion (Courtesy of Gates)

Hose Ends Detached due to Short Length: Figure 10.4 shows hose ends detached due to cutting the hose short and not respecting minimum hose length guidelines. When the hose is pressurized, it becomes even shorter causing the hose end blow off.

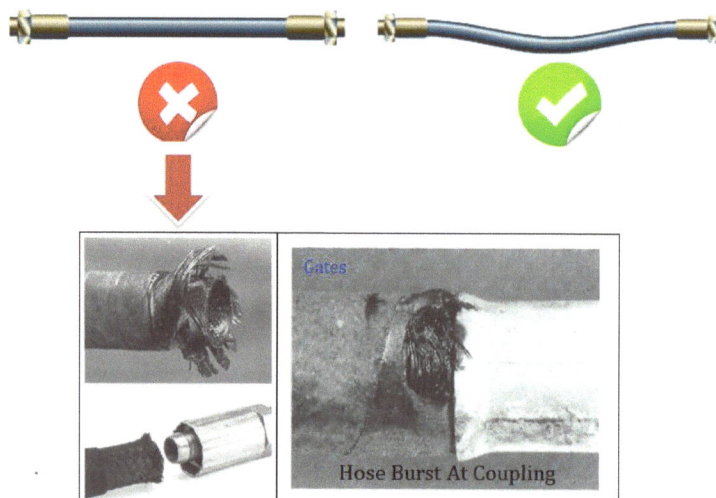

Fig. 10.4– Hose Ends Detached due to Short Length (Courtesy of Gates)

Hose Cover Cracked due to Poor Routing: Figure 10.5 shows hose cover cracked due to improper mounting and too many bends in the hose.

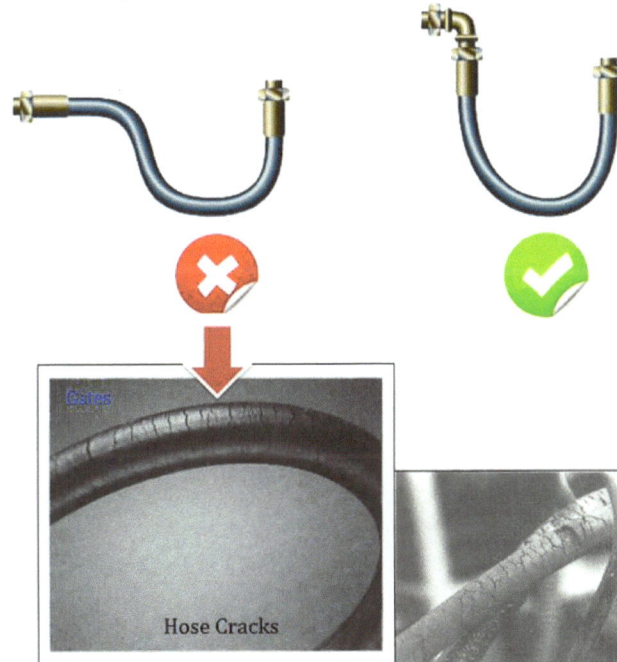

Fig. 10.5– Hose Cover Cracked due to Poor Routing (Courtesy of Gates)

Hose Plastically Deformed due to Improper Assembly: Figure 10.6 shows a hose that is twisted due to improper assembly. As a result, the hose over the time is kinked, and plastically deformed causing high pressure losses.

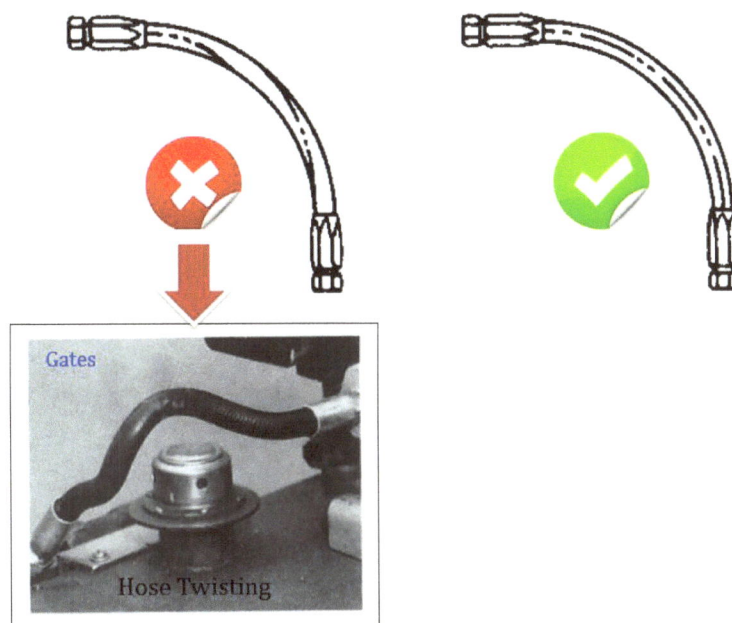

Fig. 10.6– Hose Twisting due to Improper Assembly (Courtesy of Gates)

Hose Burned due to Direct Contact with Heat Sources: Figure 10.7 shows a hose that is burned due to contact with external heat sources. It is recommended to shield hoses from such sources.

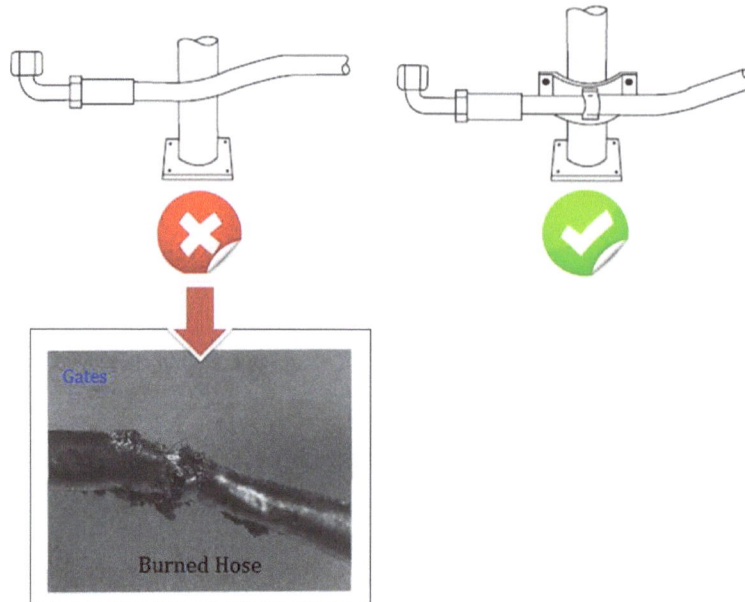

Fig. 10.7– Hose Burned due to Direct Contact with Heat Sources (Courtesy of Gates)

Hose Body Burst due to Exceeding Minimum Bend Radius: Figure 10.8 shows hose burst. The reported cause is due to the kinematical motion of the cylinder, the hose is severely bent exceeding its minimum bend radius. After number of cycles, the hose fatigued and exploded.

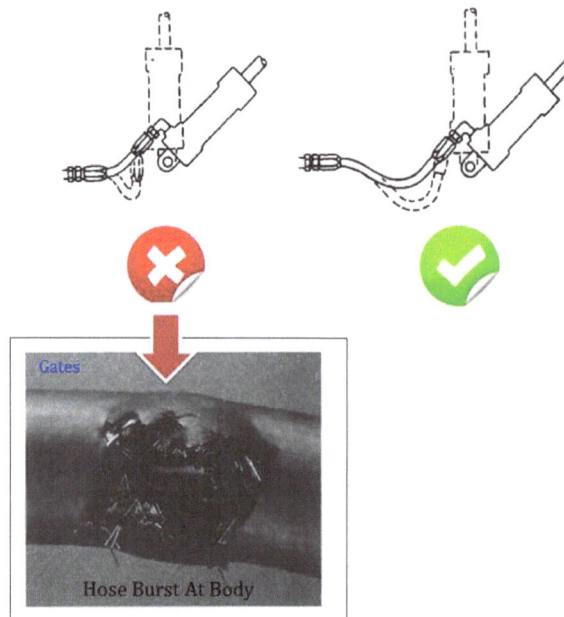

Fig. 10.8– Hose Burst due to Exceeding Minimum Bend Radius (Courtesy of Gates)

Hose Body Burst due to Exceeding lifetime: Figure 10.9 shows hose burst. The reported cause is that the hose was used for longer time than the recommended lifetime of the hose.

Fig. 10.9– Hose Body Burst due to Exceeding lifetime

Hose Failure due to Fluid Incompatibility: Figure 10.10 shows incompatible fluids causes the inner tube of the hose assembly to deteriorate, swell, and delaminate.

Fig. 10.10– Hose Failure due to Fluid Incompatibility (Courtesy of Parker)

Hose Ends Blow Off due to Short Insertion Depth: Figure 10.11 shows that, when a hose assembly is not properly assembled, it can create very dangerous situations. Fittings need to be pushed on completely to meet the recommended insertion depth. If the hose insertion depth is not met, fittings can blow off, leaving a failed hose assembly. The last grip in the fitting shell is essential to the holding strength.

Fig. 10.11– Hose Ends Blow Off due to Short Length (Courtesy of Parker)

Hose Failure due to Overheating: Hose failure can occur from overheating the hose assembly. As shown in Fig. 10.12, overheating will cause the hose to become very stiff. The inner tube will hardend and begin to crack because the plasticizers in the elastomer will break down or harden under high temperatures. In some cases, the cover may show signs of being dried out. The hose assembly may remain in its installed shape after being removed from the application and if flexed, audible cracking can be heard. In order to prevent overheated hydraulic hose assemblies, confirm hoses are rated for the temperatures required by your application. Also, reduce ambient temperatures using good ventilation or use heat guards and shields to protect the hose from nearby high-temperature areas.

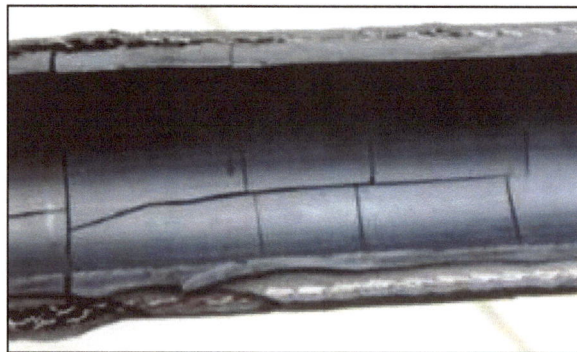

Fig. 10.12– Hose Failure due to Overheating (Courtesy of Parker)

Hose Cracked due to Exposure to Paint Spray: Fig. 10.13 shows a hose was hardened and cracked only on the side which had been exposed to paint overspray during aircraft manufacturing. The paint chemically attacked the hose. The hose hardening and cracking, and its life time is shortened to half.

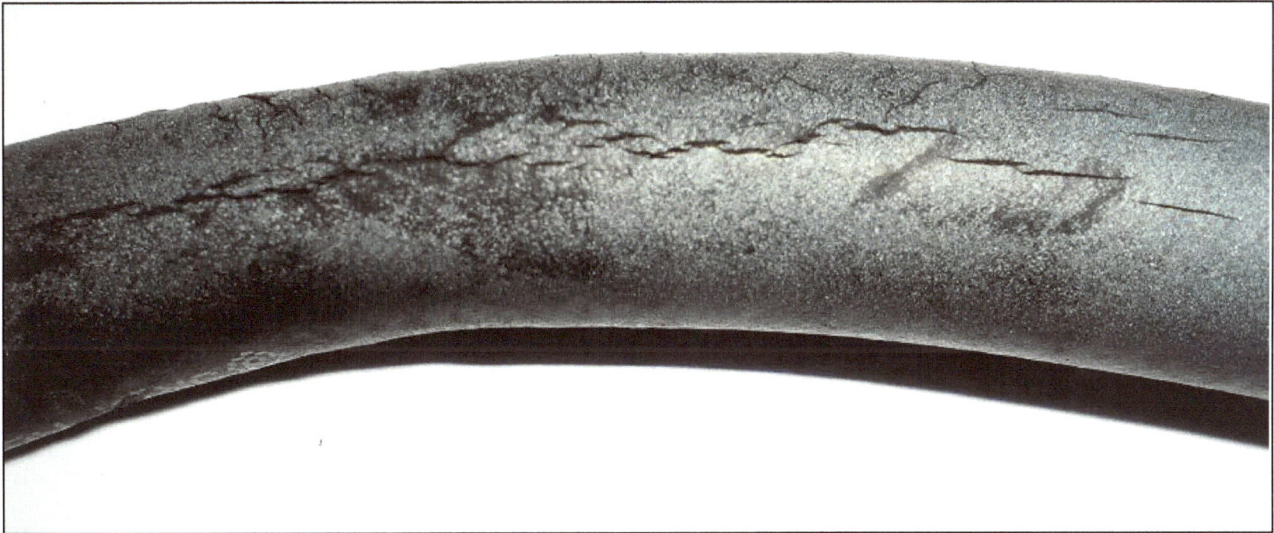

Fig. 10.13– Hose Cracked due to Exposure to Paint Spray

Tube and Pipe Burst due to Overpressure: Most people are aware that hydraulic fittings such as 37° Flare (JIC), O-Ring Face Seal, or Metric Flareless (DIN) tube fittings are designed to industry standards, such as SAE J514 or ISO 8434-1. These standards govern their dimensional characteristics. However, most of these standards also cover the performance requirements including pressure, temperature and corrosion resistance. Fig. 10.14, shows tube burst due to overpressure.

Fig. 10.14– Tube and Pipe Burst due to Overpressure (Courtesy of Parker)

Pipe and Tube Pin Holes due to Poor Material: Fig. 10.15 shows a pin hole in a hydraulic pipe due to poor material. Not any pipe can be used to transmit hydraulic energy. Selection of pipe size and material should be in accordance with appropriate industry standards for application.

Fig. 10.15– Pipe and Tube Pin Holes due to Poor Material

Pipe and Tube are Leaking due to Mechanical Stresses: Fig. 10.16 shows a leaking hydraulic pipe due to mechanical stresses as a result of the pipe weight. Such pipes must be hanged to the ceiling to release the mechanical stresses.

Fig. 10.16– Pipe and Tube are Leaking due to Mechanical Stresses

Chapter 11

Troubleshooting and Failure Analysis of Heat Exchangers

Objectives

This chapter discusses hydraulic *heat exchangers* inspection, troubleshooting, and failure analysis. In this chapter, a troubleshooting chart for heat exchanger faults is presented. This chapter also presents examples of defective heat exchangers.

The following topics are discussed in Chapter 5 in Volume 4 "Hydraulic Fluids Conditioning" of this series of textbooks:
- Contribution of Heat Exchangers:
- Air-Type versus Water-Type Oil Coolers.
- Determination of Cooling Capacity for an Oil Cooler.
- Air-Type Oil Coolers.
- Shell-and-Tube Water-Type Oil Coolers.
- Plat-Type Oil Coolers.
- Cooling-Filtration Units.
- Oil Cooling Circuit Diagram.
- Oil Temperature Automatic Control Solutions.
- Electrical Oil Heaters.

The following topics are discussed in Chapter 11 in Volume 5 "Maintenance and Safety" of this series of textbooks:
- BP-Heat Exchangers-01-Selection and Replacement.
- BP-Heat Exchangers-02-Maintenance Scheduling.
- BP-Heat Exchangers-03-Installation and Maintenance.
- BP-Heat Exchangers-04-Standard Tests and Calibration.

Brief Contents

11.1- Heat Exchangers Inspection
11.2- Heat Exchangers Troubleshooting
11.3- Heat Exchangers Failure Analysis

Chapter 11: Troubleshooting and Failure Analysis of Heat Exchangers

11.1- Heat Exchangers Inspection

Hydraulic *heat exchangers* can be air-cooled coolers, water-cooled coolers, or heaters. They can work either on/off or proportional to control working temperature within recommended limits for proper operation of hydraulic systems. Volume 4 and 5 of this series of textbooks presents the construction, operating principles, and guidelines for heat exchangers design, maintenance and safety. Table 11.1 shows a typical inspection sheet for heat exchangers.

Hydraulic Heat Exchangers Inspection Sheet	
Manufacturer	
Model #	
Serial #	
Location	
Heat Exchanger Type	☐ Air-Cooled ☐ Water-Cooled ☐ Heater
Control Mode	☐ On/Off ☐ Proportional
Heat Exchanger Cooling/Heating Capacity	
Heat Exchanger Flow	
Conditions of Heat Exchanger	

Table 11.1 – Heat Exchangers Inspection Sheet

11.2- Heat Exchangers Troubleshooting

Table 11.2 shows troubleshooting guidelines for heat exchangers.

T-Heat Exchangers-01: Heat Exchangers Troubleshooting	
Heat Exchanger is among the list of suspicious components that aren't working improperly	▪ Supply Water: ▪ Check adequate water supply flow & temperature. ▪ Check the inlet/outlet temperature of water supply. ▪ Control Parts: ▪ Check adjustment of thermostat. ▪ Check operation of thermostatic water control valve. ▪ Shell/Tube: ▪ Check if the tubes are corroded/eroded. ▪ Air-Cooled Heat Exchangers: ▪ Ambient temperatures too high. ▪ Fan speed or air flow insufficient. ▪ Check if debris has accumulated on tubes or cooling fins.

Table 11.2– Heat Exchangers Troubleshooting Chart

11.3- Heat Exchangers Failure Analysis

Tube Corrosion: The biggest threat to shell and tube heat exchangers that use carbon steel tubes is oxidation (*corrosion*) of the heat transfer surface of its tubes. The reaction between oxygen (O_2) and iron (Fe_2, Fe_3) is the most commonly observed form of corrosion. As shown in Fig. 11.1, this reaction yields a layer of iron oxide (Fe_2O_3) on carbon steel tubes which results in decreasing thermal permeation and eventually the deterioration of the tubes. This problem is difficult to overcome and is often only detected when tubes become so corroded, their thermal performance levels decrease, the fluid flow is significantly reduced, or the tubes are start leaking.

Fig. 11.1- Corroded Carbon Steel Tube (www.fluiddynamics.com)

Tube Erosion: *Erosion* of tubes is the physical wearing of the metal by fluid with extreme velocities. Water in heat exchangers erodes the tubes both internally and at the leading edges of the inlet tubes. Erosion becomes worse if the water contains silica, silt or sea water containing salt and sand. As shown in Fig. 11.2, the weakest point for tubes are the U-bends (left) due to change in direction of flow at this point. This introduces resistance causing the fluid, and any particulates in it, to impact the far wall of the tube, constantly eroding the tube at this point. The other weakest point is the inlet *tube-end (right)* where the tubes are connected through the tube sheets and face the full force of the incoming fluid.

Fig. 11.2- Heat Exchanger Tube Erosion (www.fluiddynamics.com)

Pitting of Tubes: As shown in Fig. 11.3, *Chemically-induced* corrosion can result in the pitting of heat exchanger tubes to the point where pinholes form and the tube fails and leaks. A concentrated electrochemical gradient of oxygen (O_2) and carbon dioxide (CO_2) are frequently the cause of tube wall pitting, as is the presence of excess chemical compounds such as Chloride and Sulphate often found in inadequately treated cooling water.

Fig. 11.3- Large Pitting Attack on a Copper Tube (www.fluiddynamics.com)

Thermal Fatigue: As shown in Fig. 11.4, heat exchanger tubes are vulnerable to tears and cracks due to *thermal stresses* related to constant thermal cycling or high temperature differentials. Thermal fatigue occurs when extreme temperature differences between the shell and tubes result in tube flexing.

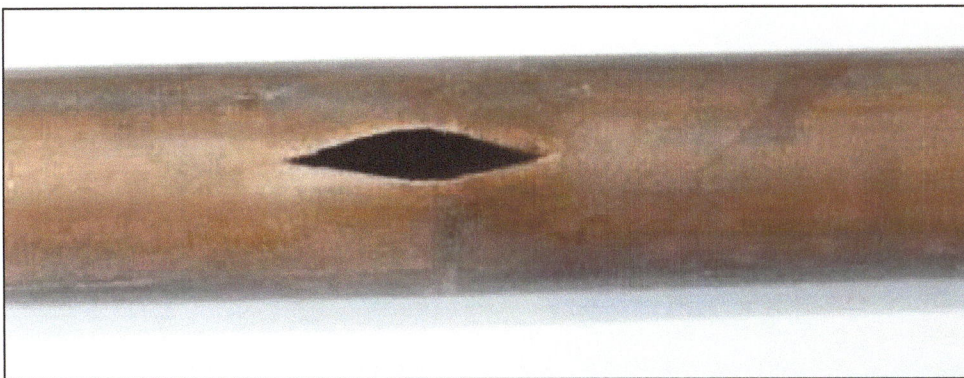

Fig. 11.4- Significant Tear in a Copper Tube due to Extreme Temperature Differences (www.fluiddynamics.com)

Ice Cold Damage: Figure 11.5 shows blown up tube stack from a Bowman water-oil cooler due to low temperature below freezing. The rig worked very well inside the mine, where the temperature was not so low. When it was taken outside for shipping, because there was water left in the coolers, water froze, expanded, and the brass pipes literally exploded from the inside out.

Fig. 11.5- Ice Cold Damage in Heat Exchangers (Courtesy of Insane Hydraulics)

Steam or Water Hammer: Steam or *water hammer* is a powerful force and can cause the rupture or collapse of either the shell or the tubes of a heat exchanger. Hammer generally occurs where there has been a surge in pressure commonly caused by a sudden interruption in cooling water flow, the rapid vaporization of stagnant water or pump malfunction. Hammer can often be heard, but only rarely will it damage the shell. As shown in Fig. 11.6, tubes are weaker than the shell and likely victims of hammer. However, damage to tubes will only be detected on internal inspection or when leaks become apparent.

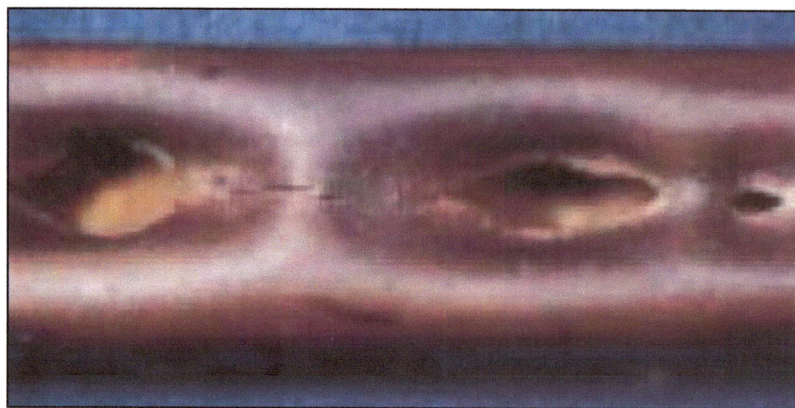

**Fig. 11.6- Collapsed and Ruptured Copper Tube Resulting from Steam Hammer
(www.fluiddynamics.com)**

Vibration/Resonance: As shown in Fig. 11.7, *vibration and resonance*, from external sources can impose powerful forces on heat exchanger tubes resulting in tubes rupturing or losing their seal with the tube-sheet and leak.

**Fig. 11.7- Broken Baffle and Worn Tubes Caused by Environmental Resonance
(www.fluiddynamics.com)**

Dusty Air-Cooled Heat Exchanger: As shown in Fig. 11.8, accumulated dust on heat exchangers significantly reduce the cooling capacity. Routine cleaning is required to maintain proper operation of the heat exchanger.

Fig. 11.8- Accumulated Dust on Air-Cooled Heat Exchanger

Chapter 12

Troubleshooting and Failure Analysis of Filters

Objectives

This chapter discusses hydraulic *filters* inspection, troubleshooting, and failure analysis. In this chapter, a troubleshooting chart for filter faults is presented. This chapter also presents examples of defective filter.

Brief Contents

12.1- Filters Inspection

12.2- Filters Troubleshooting

12.3- Filters Failure Analysis

Chapter 12: Troubleshooting and Failure Analysis of Filters

12.1- Filters Inspection

Hydraulic *filters* can be suction, pressure, or return. Filters are available in various sizes, efficiency, and dirt holding capacity. Volume 5 of this series of textbooks provides guidelines for filter maintenance and safety. Table 12.1 shows typical inspection sheet for filters.

Filters Inspection Sheet	
Manufacturer	
Model #	
Serial #	
Location	
Filter Type	☐ Suction ☐Pressure ☐Return ☐Breather
Filter Body Type	
Filter Rated Flow Rate	
Filter Beta Ratio/Efficiency	
Conditions of the Filter	

Table 12.1 – Filters Inspection Sheet

12.2- Filters Troubleshooting

Table 12.2 shows troubleshooting guidelines for filters.

T-Filters-01: Filters Troubleshooting	
• Pressure drop across the filter is larger than the rated value. • OR Clogging Indicator is activated.	• Check if the filter cartridge is clogged due to particulate contaminates, sludge, or varnish. • Check if the flow rate is above the rated value. • Check if the check valve is stuck closed.
• Media cracks. • OR Media migrates downstream the filter.	• Check if filter element is subjected to fatigue due to cyclic flow, such as when a pressure compensated pump is stroked/de-stroked very frequently.
• Improper filtration process.	• Small dirt holding capacity of the cartridge. • Bypass check valve stuck open.
• Broken filter housing.	• Too high pressure. • Shock pressure.

Table 12.2– Filters Troubleshooting Chart

12.3- Filters Failure Analysis

Filters can tell us too much about what is going on inside a machine as well as in the oil.

Filter Clogging due to Particulate Contamination: Figure 12.1 shows an example of a filter that has been clogged by dirt. The filter appears normal but the particles clogging it are smaller than the limit of vision. In operation, this filter will by-pass due to high differential pressure, thus a pressure indicator is needed to detect when the filter has reached maximum dirt holding capacity. It is advisable to unroll the pleated filter media and check what kind of material it catches. If metal flacks were found, this is an indication of machine wear. Rubber flacks indicates seal deterioration.

**Fig. 12.1- Example of Filter Blockage due to Particulate Contamination
(Courtesy of Noria Corporation)**

Filter Clogging due to Sludge: If the hydraulic fluid is exposed to high temperatures, many fluids will break-down and release resinous materials. When combined with other contaminates, sludge is formed. Sludge tends to plug small openings and orifices and interfere with heat transfer. As shown in Fig. 12.2, *Sludge* is thick polymerized compounds dissolved in warm oil. Sludge is a major source of clogging filters, strainers, and control orifices causing sudden system failure.

Fig. 12.2- Example of Filter Blockage due to Sludge

Filter Clogging due to Varnish: *Varnish* is a product of chemical degradation of hydraulic fluids. As shown in Fig. 12.3, varnish clogs filters. Furthermore, varnish acts as an insulator reducing the effect of heat exchangers.

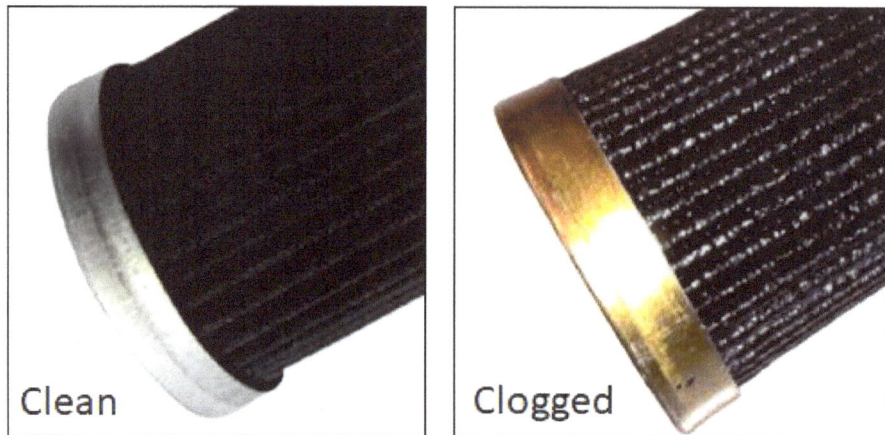

Clean   Clogged

Fig. 12.3- Example of Filter Blockage due to Varnish

Filter Media Collapse due to Cyclic Flow: As shown in Fig. 12.4, *cyclic flow* can cause fatigue of filter element structure and result in cracking of the pleats unless proper filter medium support is included within the element. Surge or cyclic flow occurs in cases such as when a pressure compensated pump is stroked/de-stroked very frequently.

Fig. 12.4- Example of Filter Media Collapse due to Cyclic or Surge Flow

Chapter 13
Hydraulic Systems Troubleshooting

Objectives

This chapter introduces troubleshooting charts for failures of generic hydraulic systems. Each troubleshooting chart includes relevant notes and examples for better understanding.

Brief Contents

Chapter 13: Hydraulic Systems Troubleshooting

13.1-Features of Hydraulic Systems Failures

Unlike other power transmission and control systems, tracking the root cause of hydraulic system failures is challenging because:

- **Closed System:** Can't see inside the system.
- **Common Symptoms:** Same symptoms could be due to different causes.
- **Failures are Transferrable:** Once a problem occurred in one place, it moves forward and transfers with the fluid to other components, e.g. wear products and overheated fluid pass through downstream components.
- **Failures are Accelerated:** The rate of growth isn't linear, e.g. leakage is proportional to the cube of clearance.
- **Chain Action (see Fig. 13.1):** Failures feed each other in a chain action, e.g. (internal leakage – heat – poor viscosity – less lubrication – wear – internal leakage).

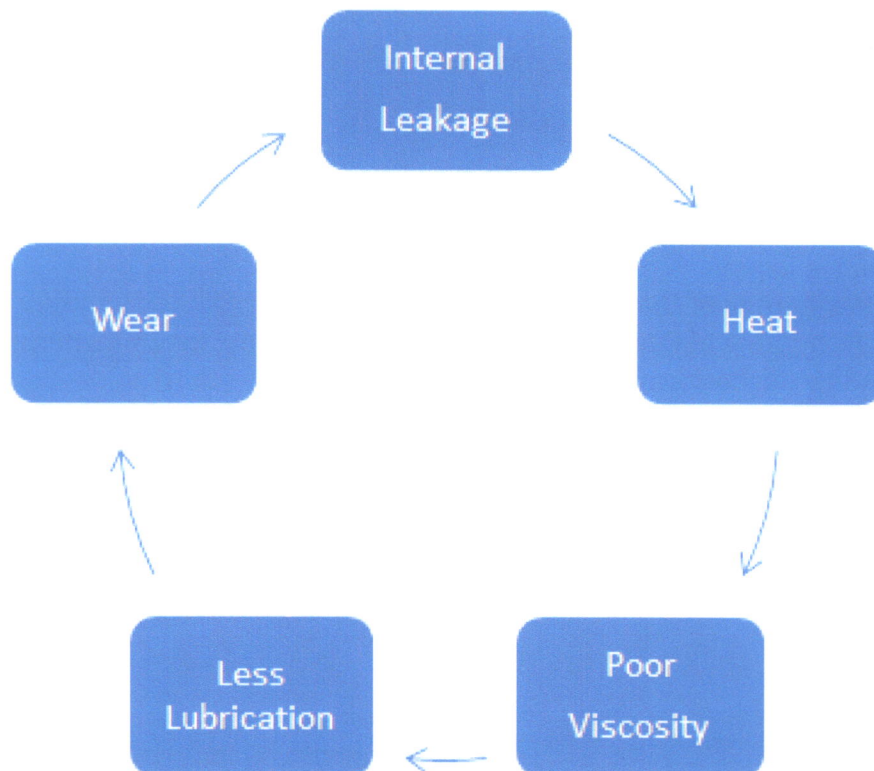

Fig. 13.1- Chain Action of Hydraulic System Failures

13.2-Main Causes of Hydraulic Systems Failures

Some failures of hydraulic systems occur due to specific reasons tied to specific application or operating conditions. In addition to that, there are common failures that occur for any hydraulic system due to common reasons. Figure 13.2 shows how hydraulic systems failures are categorized. The following subsections provides examples of each category.

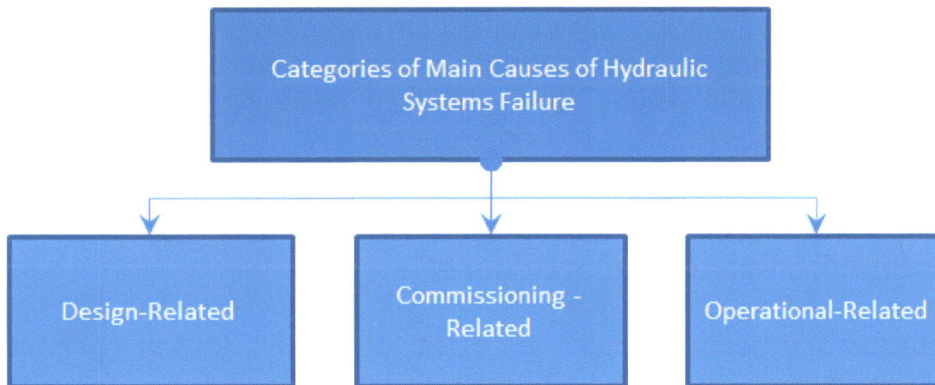

Fig. 13.2- Categories of Main Causes of Hydraulic System Failures

13.2.1-Design-Related Failure Causes

Poorly designed systems will soon become source of troubles. <u>For example:</u>

- **Incorrect Component Sizing:** Proper sizing of hydraulic components is the most important design step: Oversizing hydraulic valves result in losing the controllability, and under-sizing valves result in heat generation.
- **Lack of Predicting System Performance:** System designers should be capable of predicting the system performance and consider the proper design solutions accordingly. For example, if a hydraulic system frequently experiences pressure spikes, an accumulator may be used to reduce such spikes.
- **Specifying Wrong Fluid:** The oil is the most important component of any hydraulic system because it affects both machine performance and service life. For example, using classical mineral fluid in a high-temperature application such as steel mills may not function properly. Using oil with low viscosity index in aerospace applications is another bad example.
- **Poor Filtration System Locations:** Designing a filtration system is a process that contains several elements such as filter location, filter dirt holding capacity, filter Beta ratio, etc. Mistakes in filtration system design is a direct source of future problems. An example of that is placing a filter on the pump inlet without reviewing the acceptable vacuum conditions at the pump inlet.
- **Poor Reservoir Design:** A poorly designed reservoir could be a source of hydraulic fluid aeration.

13.2.2-Commissioning-Related Failure Causes

Proper system design isn't the end of story, if a hydraulic system isn't properly commissioned, failures may arise. Therefore, it always advisable to review the assembly instructions provided by the manufacturer. <u>For example:</u>

- Improper routing of hydraulic lines results in lots of line losses.
- Improper assembly of vanes in a vane pump results in loss of flow.
- Improper assembly of a pump may result in pump cavitation.
- Shaft misalignment in pumps and motors results in vibration and shaft seal failure.
- Believing hydraulic pumps are self-primed is wrong assumption that may lead to premature failure.

13.2.3-Operationl-Related Failure Causes

A well-designed and commissioned system can still fail if it isn't operated or maintained properly. <u>For example:</u>

- **Lack of Understanding:** Misunderstanding how the hydraulic system operates may result in failures. For example, changing a location of a hydraulic valve may change the sequence of operation of a system or damage hydraulic actuators. Therefore, training is very important to improve the awareness of system operation and possible failures.
- **Lack of Maintenance:** Improper or lack of maintenance is a main cause of system failures. Well-maintained hydraulic systems can last for 25 years with no failures.
- **Low Fluid Volume:** Proper oil volume must be maintained. Otherwise, the pump may cavitate or air entrain will cause actuators to move erratically.
- **Changing Fluid Timely:** Hydraulic fluid is the life blood of a machine, and it reflects its health. Therefore, hydraulic fluid must be changed in case of fluid degradation of the base oil or depletion of the additive package. Continuing to operate with the base oil degraded or additives depleted, compromises the service life of every other component in the hydraulic system.
- **Keeping working temperature within allowable limits (low – high):** Operating temperature affects physical and chemical properties of the fluid. Both running too hot or too cold result in system problems. But the question is How hot is too hot and how cold is too cold for a hydraulic system? It depends mainly on the viscosity and viscosity index (rate of change in viscosity with temperature) of the oil, and the type of hydraulic components in the system. Therefore, hydraulic fluids operating temperature must be maintained within the allowable limits (minimum and maximum) that are specified by the system manufacturer.
- **Fluid Contamination:** Fluid contamination is the main cause of hydraulic system failure. Fluid contamination has several sources, forms and symptoms. All must be monitored on a continuous base for improving hydraulic systems reliability.
- **Misuse:** Abusing hydraulic systems cause failure. For example, over pressurized cylinder may cause seal failure, and over speeding motor may result in premature failure.

13.3- Fluid Aeration

Hydraulic fluid is contaminated by air from different sources as follows:

Air Leaks into System: As shown in Fig. 13.3, air can find its way into the system due to insufficient oil volume or leak in suction side of the pump.

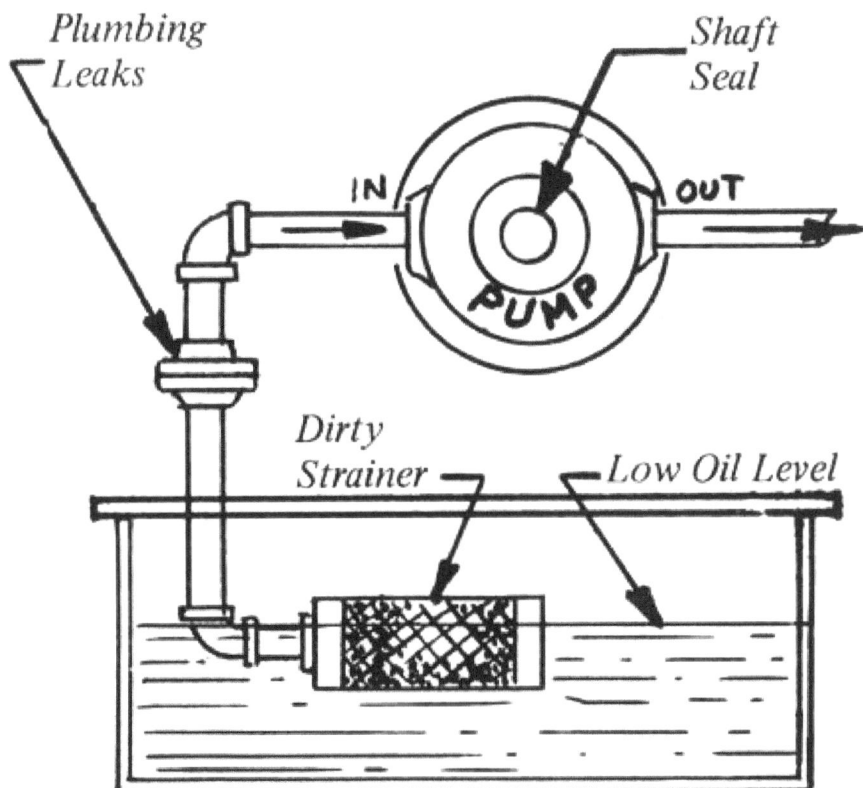

Fig. 13.3- Sources of Fluid Gaseous Contamination (Courtesy of Womack)

Air Separation from Fluid: Hydraulic fluids contain (7-10) % by volume air. This amount of air, under ambient conditions, is homogeneously dissolved within the fluid on the molecular level. This amount of dissolved air does not affect the fluid properties or system performance. When the fluid passes through a negative pressure zone, dissolved air separates from the fluid in form of bubbles.

Hydraulic Fluid Evaporation: If the hydraulic fluid is subjected to severe temperature and vacuum conditions, the fluid itself evaporates forming bubbles.

As shown in Fig. 13.4, fluid appearance due to existence of air in oil changes from cloudy to creamy color depending on the amount and the size of air bubbles.

Aeration: Sustained tiny emulsified bubbles below the surface of the fluid is called aeration. Air emulsifies with the oil and causes it to have a milky appearance, but the oil will usually become clear in about an hour after shut-down.

Foaming: Accumulation of bubbles on top of the fluid surface.

Result of Aeration: As a result of fluid *Aeration*, the following consequences occur:
- Power Loss.
- Noisy and vibrated operation.
- Poor lubrication.
- Increased rate of oxidation.
- Erratic/Sluggish machine motion and control.
- Drastic reduction in fluid's bulk Modulus.

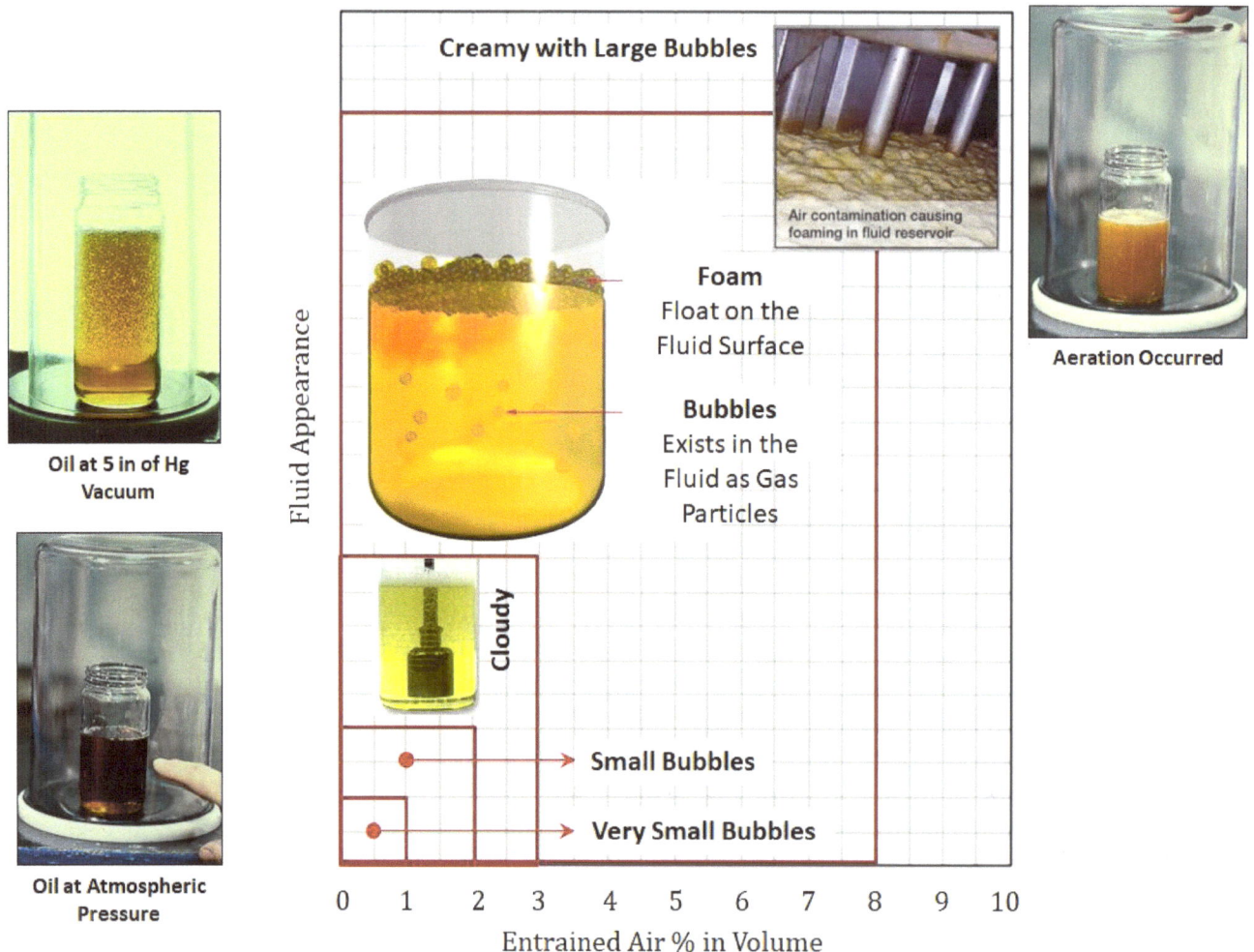

Fig. 13.4- Foam and Entrainment (Courtesy of Noria Corporation)

As shown in Fig. 13.5 and Table 13.1, hydraulic fluids have different abilities to suppress aeration. The figure shows that less viscous fluids have better aeration suppression. The figure shows also that larger bubbles rise faster to the fluid surface and dissipate faster.

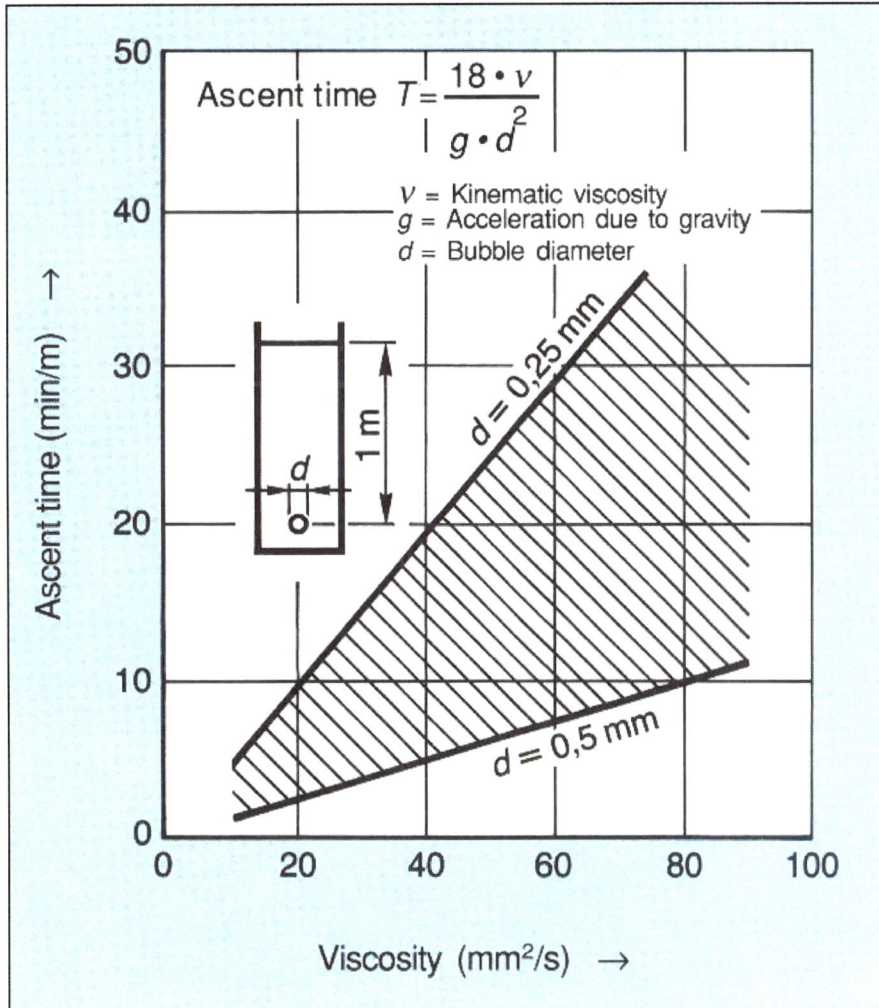

Fig. 13.5- **Hydraulic Fluids Foam Suppression-Ability (Courtesy of Bosch Rexroth)**

Viscosity	Release Time
ISO VG 10, 22, and 32	Maximum 5 min
ISO VG 46 and 68	Maximum 10 min
ISO VG 100	Maximum 14 min

Table 13.1- **Air Separation Capacity in Minutes at 50 °C (Courtesy of Bosch Rexroth)**

Table 13.2 shows the relevant troubleshooting chart.

T-System-01-Fluid Aeration	
Reservoir	
Reservoir fluid level too low?	▪ Follow the guidelines to make up the oil in the reservoir to the specified level.
Poor reservoir design? Return lines were improperly located?	▪ Review reservoir design about air removal. ▪ Consider using screen separator placed on 45-degree angle between suction and return side. ▪ Consider using baffle plate between suction and return lines to elongate oil residence time in the reservoir. ▪ Consider return line discharges below the oil surface to reduce fluid vertexing and sloshing.
Transmission Lines	
Fluid returns at high speed?	▪ Resize the return line so that the return fluid speed is (1.2 - 2.1) m/s = (4 - 7) ft/s. Using return diffuser helps getting rid of air.
Turbulent flow in the system?	▪ Review system design to secure laminar flow (flow rate – conductor size – fluid viscosity).
Air Leaking	
Air Leaks into the Pump?	▪ Consult Chart: ▪ **"T-Pump-11-Air Leaks into the Pump"**
Air Leaks into the cylinder?	▪ Check if air leaking into cylinder through a rod seal when an overrunning load drives the cylinder developing vacuum.
Gas leaks from the accumulator?	▪ Consult Chart: ▪ **"T-Accumulator-01-Accumulator Troubleshooting"**
Hydraulic Fluids	
Foam suppression additives are depleted.	▪ Consult the fluid supplier about the best practices for adding foam suppressors to the fluid.

Table 13.2- Troubleshooting Chart (T-System-01-Fluid Aeration)

13.4- Pump Cavitation

Cavitation is always been confused with aeration. Cavitation in simplest definition is that the pump is starving for fluid and the fluid isn't fully filling the cavities of the pumping chambers. Cavitation occur due several reasons. However, the cavitation mechanism is simply formation of air bubbles in the suctions side of the pump and they collapse in the pressure side of the pump. Formation of bubbles within the liquid begin when the liquid pressure become equal or even close to the vapor pressure of the fluid at a given temperature. Bubbles increase rapidly in size and in numbers as shown in Fig. 13.6.

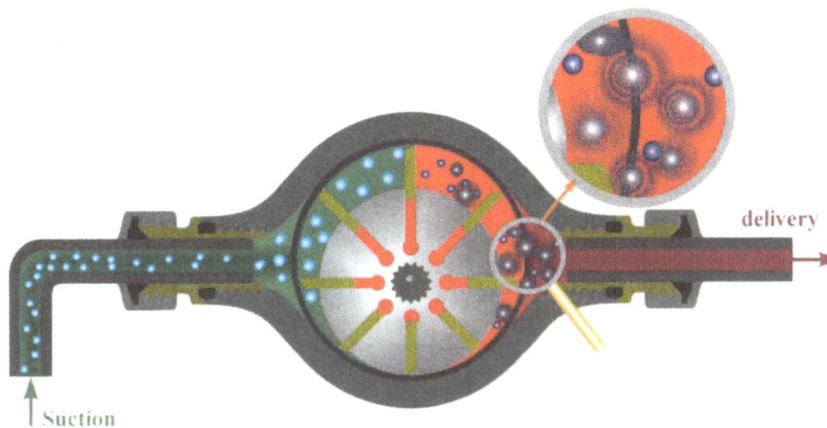

Fig. 13.6- Pump Destruction due to Cavitation (Courtesy of Assofluid)

Subsequently as shown in Fig. 13.7, when the bubbles enter a zone of high pressure, they are super-compressed (imploded) with microjet action.

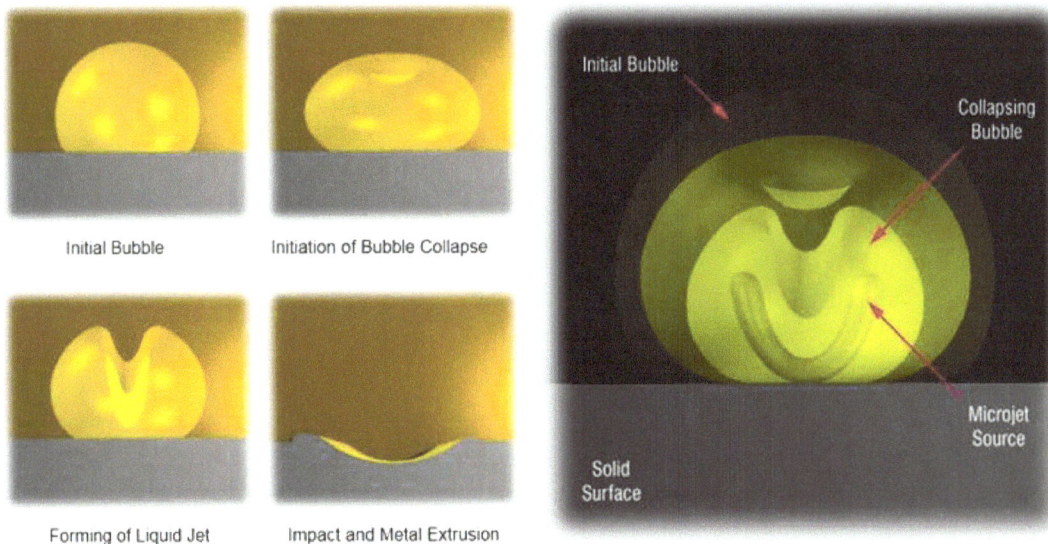

Fig. 13.7- Pump Cavitation (Courtesy of Noria)

As a result of pump *Cavitation,* in addition to consequences of aeration, the following consequences occur:

- Implosion of bubbles is accompanied by a microjet shock load, unrepairable destruction of material bonds, sound emission, pump vibration, and overheating. As shown in Fig. 13.8, different metals have different resistance to cavitation.
- Loss of pump flow and pressure.
- As shown in the figure, dieseling effect on seals leaving burning spots.

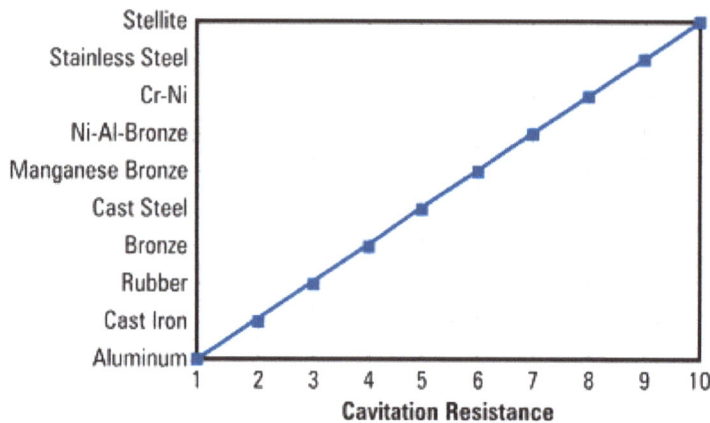

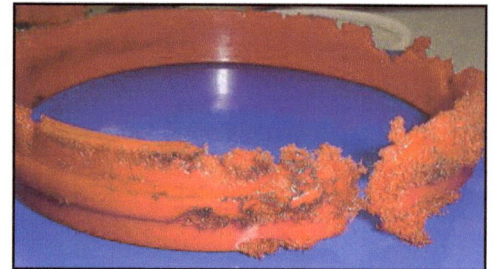

Fig. 13.8- Cavitation Resistance (Courtesy of Noria)

In order to avoid aeration and cavitation, the following actions must be considered:

❖ **System Design Considerations:**
- **Oil Reservoir Design:** Design reservoir to help remove bubbles and reduce turbulence.
- **Boosting Pump Intake:** Boost intake pressure of large size and high-speed pumps.
- **Vacuum Switch:** Use a vacuum switch as a preventive solution to stop the pump in case if pressure at the pump intake is reduced below recommended.

❖ **Pump Installation Considerations:**
- **Pump Placement:** Place pump below fluid surface to build positive intake pressure.
- **Intake Line:** Properly size and install intake line as per the relevant guideline (Review Chapter 2 in Volume 4 "Hydraulic Fluids Conditioning" of this series of textbooks).
- **Pump Location:** Shouldn't be installed above the reservoir with too high suction head.
- **Pump Priming:** Review manufacturer instructions for proper priming.
- **Suction Strainer:** Consult the pump's datasheet to define the minimum mesh size and the size of the strainer in relation to the pump flow rate.

❖ **System Operation Considerations:**
- **Working Temperature:** keep within recommended range.
- **Hydraulic Fluid:** use the fluid with recommended viscosity.
- **Driving Speed:** keep it within recommended range.
- **Suction Strainer/Suction Line:** inspect and clean periodically.
- **Oil Level:** inspect periodically.

Table 13.3 shows the relevant troubleshooting chart.

T-System-02-Pump Cavitation	
Reservoir: Fluid level is too low, reservoir with no baffle, or too shallow?	▪ Follow the guidelines for reservoir design and make up the oil to the specified maximum level.
Reservoir isn't vented?	▪ Check clogged air breather in open tanks. ▪ Check pressure of closed reservoir.
Suction Line: Suction valve is partially or fully closed?	▪ Fully open the suction valve. ▪ Lock the suction valve in fully opened position.
Suction filter/strainer/air breather are clogged or undersized?	▪ Wash or replace strainer or suction filter. ▪ Check manufacturer recommendation about using suction strainer for this pump.
Suction line is undersized, kinked, restricted, sucking air, or plugged.	▪ Check intake line size vs. pump intake port. ▪ Flow speed = (0.6 - 1.2) m/s = (2 - 4) ft/s.
Suction line is too long, too short, or has too many bends?	▪ Review design of intake line (Chapter 2 – Volume 4)
Is the suction line flexible hose?	▪ Make sure it is a suction hose (not a pressure hose) because inner layers of pressure hoses are not made to carry negative pressure.
Hydraulic Fluid: Fluid is too hot?	Consult Chart: **"T-System-04-Excessive System Heat"**
Fluid is too cold or too viscous?	▪ Review the pour point of the fluid. ▪ Warm the fluid 10 degrees C above the pour point before starting the machine. ▪ Review manufacturers recommendation about the fluid viscosity and replace the fluid if needed.
Pump: Pump is placed too high from fluid surface or in a wrong orientation?	▪ Review the manufacturer's instructions about the pump maximum suction head or pump orientation (review Volume 5).
Does the pump rotate at high speed?	▪ Review the manufacturer's instructions about the maximum allowable speed.
Is the pump supercharged?	Check the pump inlet pressure and the inspect the boosting pump.

Table 13.3- Troubleshooting Chart (T-System-02-Pump Cavitation)

13.5- Excessive System Noise & Vibration

Vibration in a hydraulic system gradually loosens the fittings and causes fatigue failure of the system and transmission lines.

Noise in a hydraulic system causes environmental pollution and long-term personal disability. Noise can be structure-borne, fluid-borne and air-borne noise.

Table 13.4 shows the permissible noise exposure level based on standards provided by **OSHA Standard (Act of 1970).** If the sound level cannot be reduced below the allowable maximum, ear protection must be provided in the workspace.

Table 13.5 shows the relevant troubleshooting chart.

Hours/Day	Sound Level (db.)
8	90
6	92
4	95
3	97
2	100
1-1.5	102
1	105
0.5	110
0.25 or less	115

Table 13.4– Permissible Noise Exposure

T-System-03-Excessive System Noise & Vibration	
Hydraulic Fluid:	
Fluid Aeration (Is there air in the system or fluid looks milky)?	▪ Consult Chart: ▪ **"T-System-01-Fluid Aeration".**
Machine is too cold, or fluid viscosity is too high?	▪ Check fluid viscosity and resolve accordingly. ▪ Check recommended temperature and resolve accordingly.
Transmission Lines:	
Is the pipework adequately supported?	▪ Support the pipework as per the installation guidelines.
Pump:	
Pump is noticeably noisy and vibrating?	▪ Consult Charts: ▪ **"T-Pump-12-Excessive Pump Noise and Vibration".**
Other Units:	
Electrohydraulic valve shuddering?	▪ Defective solenoid or voltage too low. ▪ Large flow and flow forces through the valve. ▪ Improper control settings.
Is the noise associated to a specific unit?	▪ Consult Chart ▪ **"T-Unit-02-Noisy Unit".**

Table 13.5- Troubleshooting Chart (T-System-03-Excessive System Noise & Vibration)

13.6- Excessive System Heat

There are two ways to solve overheated hydraulic systems: decrease the heat generation or increase the heat dissipation. By calculating the system's heat load and knowing the cooling capacity, you can determine if the system is stable or unstable. If the system is unstable, you must decide what can be done to remedy the problem. As mentioned, decreasing the heat load is one option, but this isn't always possible. The other option is to increase cooling. This can be accomplished by using larger heat exchangers. But, resolving the root cause of heat generation in the system would be more wise decision. Therefore, knowing the sources of generating heat and consequences of overheating hydraulic systems offers road map for troubleshooting.

Sources of Heat in Hydraulic Systems: There are various sources that contribute to adding *heat* to hydraulic fluids. These sources can be broadly classified as Design-Related and Operation-Related. Figures 13.9 and 13.10 shows examples of design-related an operation-related sources for overheating a hydraulic system. Approximately 25 percent of the input electrical horsepower will be used to overcome heat losses in the system. Operator habits could cause system overheating. For example, an operator drives the system against blocked load making the pump flow go over the relief valve frequently unnecessarily.

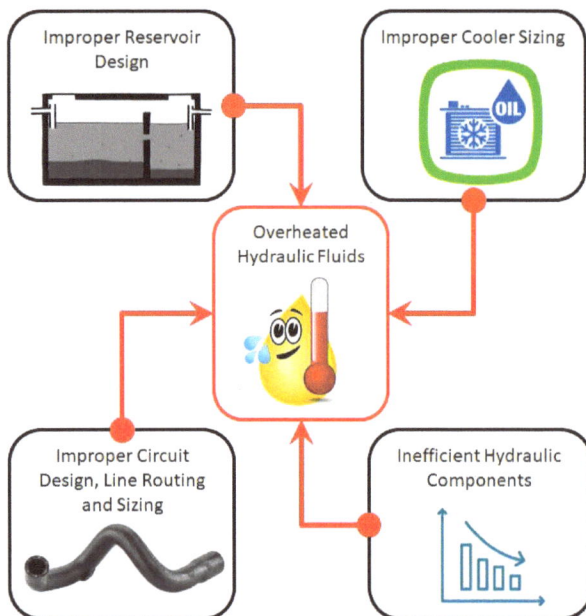

Fig. 13.9- Design-Related Heat Sources

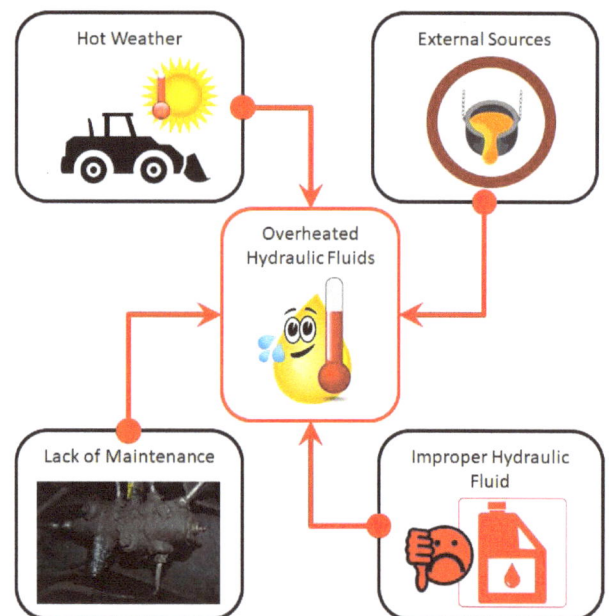

Fig. 13.10- Operation-Related Heat Sources

Consequences of Hydraulic System Overheating: Heat contamination reduces oil viscosity, which in turn reduces the fluid's ability to lubricate components. This thinning of the oil causes increased metal-to-metal friction. Heat contamination also changes the chemical properties of hydraulic oil. It can reduce the additives in the hydraulic oils, increase sludge and also accelerate oxidation.

❖ **Petroleum-Based Fluids:** Most fluid manufacturers specify optimum range of working temperature for their products, typically from 38°C to 54°C (100°F to 130°F) even though many fluids are operated above this temperature range. **Typical critical working temperature is 70°C (158°F). Every incremental increase of 10°C (18°F) higher than the critical temperature doubles the oxidation rate and life of the oil is cut in half. For example, running a system consistently at 80°C (176°F) would reduce the fluid life by 75%. Working at a temperature above 82°C (180°F), damages most seal compounds and accelerates oil degradation.**

❖ **Water-Based Fluids:** Overheating causes water evaporation, changes the ratio of water to base fluid, increases both the viscosity and additive concentration, and reduces the fluids fire resistance.

Table 13.6 shows the relevant troubleshooting chart.

T-System-04-Excessive System Heat	
Oil Cooler: Insufficient cooling capacity?	Check: ▪ Cooling water supply temperature and valve. ▪ Thermostat adjustment/operation. ▪ Cleanliness of the cooler (air/water). ▪ Dirt accumulated on top of outer surfaces of reservoir, lines, and other components.
Reservoir: Reservoir fluid level too low?	▪ Follow the guidelines to make up the oil in the reservoir to the specified level.
Poor reservoir placement?	▪ Avoid placing reservoir in a point of less air flow. ▪ Use forced air ventilation (e.g. industrial fan) to drive airflow around the reservoir. ▪ Temperature at the exterior of the reservoir should not exceed 140ºF (60ºC). ▪ Exterior of reservoir and all components must be kept clean to ensure that no hot spots develop as a result of accumulated dust and dirt.
Poor reservoir design?	▪ Review guidelines of placing suction line and return line. ▪ Use of baffle plate to separate suction line from return line. ▪ Reservoir size it too small.
Hydraulic Fluid: Improper fluid conditions?	▪ Check fluid viscosity, cleanliness, and aeration.

Working Pressure: System pressure above normal?	▪ Check if load on actuators above normal. ▪ Reset maximum system pressure if possible.
Working Flow: Excessive flow over relief valve?	▪ Check: relief valve setting, pump driving speed, maximum flow of variable displacement pump, and system duty cycle. ▪ Pump may not be unloaded between cycles. ▪ Compensator of pressure-compensated pump may be set higher than the relief valve.
Transmission Lines: Undersized transmission lines?	▪ Undersized transmission lines cause turbulent flow and increase wasted energy and heat generation.
Flow Control Valves:	▪ Check if the valve is undersized. ▪ Check if needle valve is installed backward.
Directional Control Valves:	▪ Check if the valve is undersized.
Pressure Relief Valve	▪ Check if the valve is misadjusted, so it is leaking oil to the tank.
Pressure Compensated Pump Pressure compensator is misadjusted.	▪ Check and readjust.
Other Units: Any sign of worn component and/or internal leakage?	▪ Examine and test valves, cylinders, motors, etc. for internal leaks. ▪ If wear is abnormal, replace the component at fault.
Heat is associated with a specific component?	▪ Consult Chart: ▪ **"T-Unit-03-Excessively Hot Unit".**

Table 13.6- Troubleshooting Chart (T-System-04-Excessive System Heat)

13.7- Low Power System

As shown in Fig. 13.11, when a machine is barely carrying a load, this means that there is something wrong. The sign for reduced load carrying capacity is the actuator moves when the load is reduced. Table 13.7 shows the relevant troubleshooting chart.

Figure 13.11- Low Power System

T-System-05-Low Power System	
Engine & Power Transmission: Insufficient Power and Torque at the Power Take-offs?	▪ Air filter of the engine is clogged. ▪ Power transmission defective (e.g. V-belt or toothed belt slippage, key sheared off at pump-motor coupling). ▪ Check reference duty cycle of the machine.
Pump: Pump outlet low pressure?	▪ Consult Chart: ▪ **"T-Pump-06-Low Pressure at the Pump Outlet".**
Actuator: Actuator Leaks?	▪ Consult Chart: ▪ **"T-Cylinder-1-Cylinder Troubleshooting".**
DCV: DCV at Fault?	▪ Consult Chart: ▪ **"T-Valve-01-DCV Troubleshooting".**
PRV: PRV at Fault?	▪ Consult Chart: ▪ **"T-Valve-03-PRV Troubleshooting".**
Filter: Pressure filter blocked?	▪ Check and act accordingly.

Table 13.7- Troubleshooting Chart (T-System-05-Low Power System)

13.8- Faulty System Sequence

As shown in Fig. 13.12, some hydraulic control systems follow a certain sequence. If this sequence is mistakenly changed, the system may not operate properly causing actuators to move against each other. Table 13.8 shows the relevant troubleshooting chart.

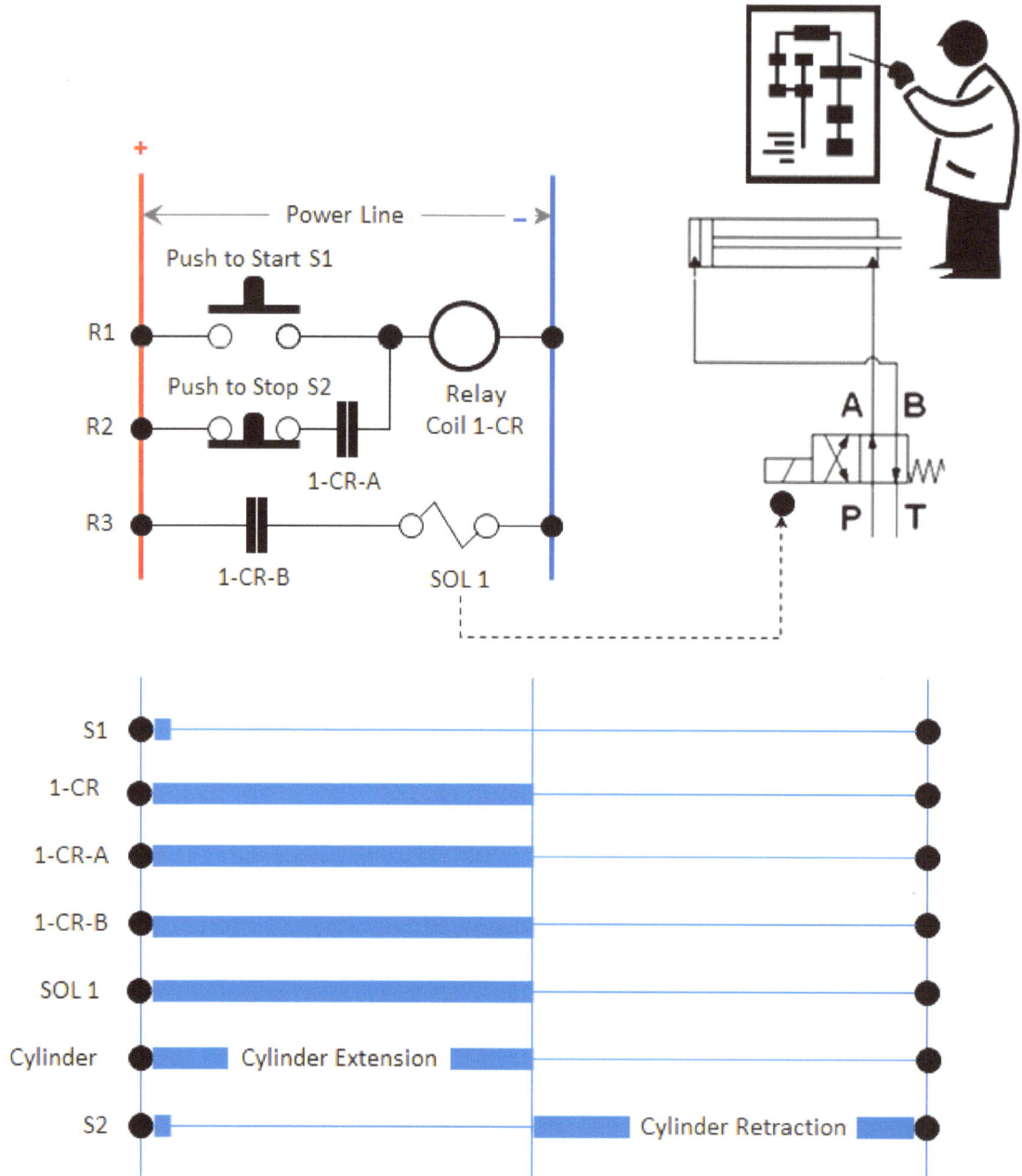

Figure 13.12- Hydraulic Control Sequence Diagram

T-System-06-Faulty System Sequence	
Actuator: Actuator Slow Performance?	▪ Consult Chart: ▪ **"T-System-10-Actuator Slow Performance".**
Actuator Fast Performance?	▪ Consult Chart: ▪ **"T-System-11-Actuator Fast Performance".**
Actuator Erratic Performance?	▪ Consult Chart: ▪ **"T-System-12-Actuator Erratic Performance".**
Actuator Moves in Wrong Direction?	▪ Consult Chart: ▪ **"T-System-13-Actuator Moves in Wrong Direction".**
Actuator Stops to Move?	▪ Consult Chart: ▪ **"T-System-14-Actuator Stops to Move".**
DCV: DCV at fault?	▪ Consult Chart: ▪ **"T-Valve-01-DCV Troubleshooting".**
Controller: System controlled by a PLC or host controller?	▪ Check control code. ▪ Check digital/analog signals out of the controller.
System contains sensors	▪ Check wiring of sensors. ▪ Check proper functioning of sensors.

Table 13.8- Troubleshooting Chart (T-System-06-Faulty System Sequence)

13.9- External Leakage

As shown in Fig. 13.13, external leakage causes the following problems:

1. Cost of making up the lost fluid.
2. Cost of removing the resulting environmental pollution.
3. Risk of personal safety due to possible slipping.
4. Risk of personal safety due to possible fire hazard.
5. Loss of system pressure if a pressure line is leaking
6. Loss of actuator power if cylinder seal is leaking
7. Increasing working temperature due to increased fluid circulation and possibilities of fluid aeration and pump cavitation.
8. Loss of fluid increases contamination due to adding contaminates with new fluid and improper practices when adding fluid to the reservoir.

Table 13.9 shows the relevant troubleshooting chart.

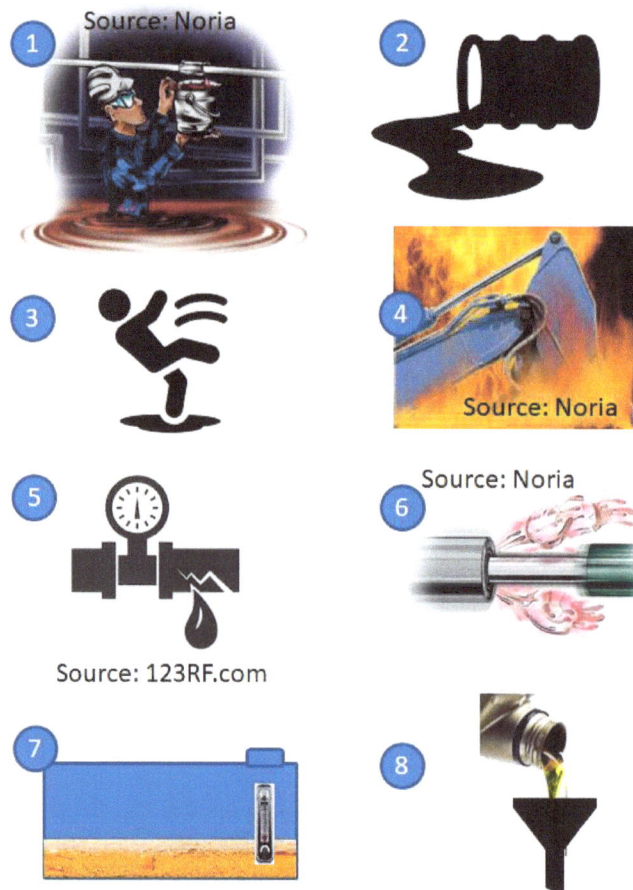

Figure 13.13- Examples of External Leakage

T-System-07-External Leakage	
Working Pressure: System pressure is too high?	▪ Review working pressure and resolve problem accordingly.
Working Temperature: Excessively hot system?	▪ Check the deterioration of elastomeric seals because of elevated fluid temperatures.
Transmission Lines: Leakage is traced to a certain fitting or a conductor?	▪ Check if a fitting is loose or over tightened. ▪ Check if different brands of fittings are mixed. ▪ Check if a cutting ring or sealing element is reused. ▪ Check if a conductor is damaged. ▪ Check if a conductor is improperly clamped. ▪ Check if a conductor is improper installed leaving mechanical stress in it. ▪ Check if a tube is properly flared.
Cylinder: Leaking Cylinder?	▪ Consult Chart: ▪ **"T-System-15-Actuator Load Drifting".**
Hydraulic Fluids: Incompatible fluid?	▪ Check the compatibility of the fluid with the shaft seals of pumps, motors, and cylinders.
Other Units: Leakage is traced to one unit.	Check the seal condition of the leaking component.

Table 13.9- Troubleshooting Chart (T-System-07-External Leakage)

13.10- Troubleshooting of Open Hydraulic Circuits

An *Open Hydraulic Circuit* is the one that has the return oil from the system going directly to the reservoir. Figure 13.14 shows the common components in a typical open circuit diagram. Recognize points A, B, and C on the figure. They will be used in the troubleshooting process.

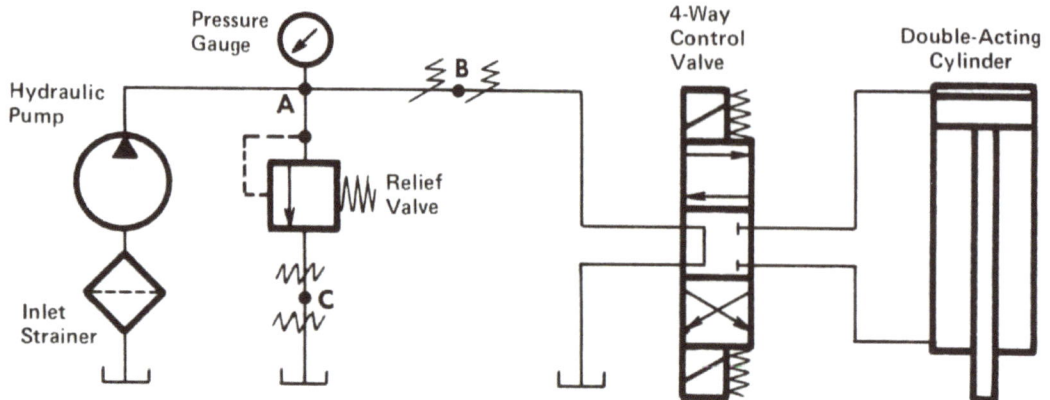

Figure 13.14- Typical Circuit Diagram of Open Hydraulic Circuit (Courtesy of Womack)

Table 13.10 shows the relevant troubleshooting chart.

T-System-08-Troubleshooting of Open Hydraulic Circuits	
Improper operation of an open circuit	▪ Step 1: Check Pump Inlet Strainer ▪ Step 2: Check Pump and Relief Valve: ▪ Step 3: Check Pump: ▪ Step 4: Check Pressure Relief Valve: ▪ Step 5: Check Cylinder: ▪ Step 6: Check Directional Control Valve:

Table 13.10- Troubleshooting Chart (T-System-08-Troubleshooting of Open Hydraulic Circuits)

The following section presents a step-by-step troubleshooting procedure for an open circuit system. However, the troubleshooter has the right to skip any step based on their understanding of the system conditions or the original source of the fault.

❖ **Step 1: Pump Inlet Strainer:**
  ▪ **Fault:** A dirty *strainer* is a major cause of pump cavitation and permanent failure.
  ▪ **Location:** Commonly It is located inside the reservoir under the oil level (See Fig. 13.15). Sometimes it can be located outside the reservoir on the intake line before the pump.
  ▪ **Replacement:** A strainer should be replaced in case if (See Fig. 13.16).
    o There are holes in the mesh or has physical damage.
    o There are heavy spots of varnish.
    o Unwastable strainers.

- **Routine:** Pump strainer should be cleaned routinely, no matter if it looks clean or dirty.
- **Air Blow:** Wire mesh strainers are cleaned with blowing air from the inside out.
- **If Varnish Found:** If there is any deposit of a brown varnish on the strainer, it should be washed in a solvent, scrubbing with a bristle brush.
- **Solvent:** It should be compatible with the fluid in the system. For example, kerosene can be used with mineral oils. On systems using synthetic or fire-resistant fluids, use some of the same fluid for cleaning the strainer.
- **Caution!** Do not use solvents like gasoline, thinner, etc., which are explosive, highly flammable, and contains no lubricant.
- **Re-installing:** Before reinstalling the strainer, inspect all joints in the inlet plumbing for air leaks, particularly at union joints.

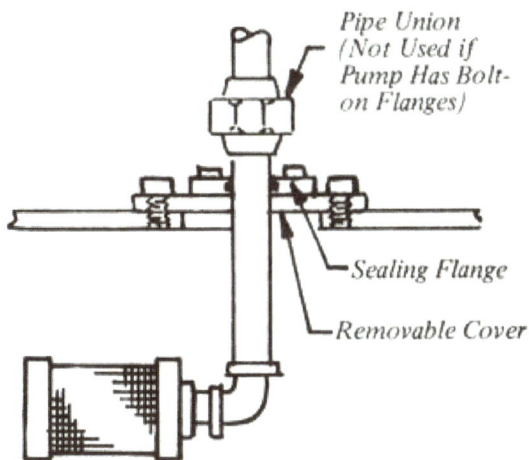

Uncouple inlet line, remove cover plate, and withdraw strainer from reservoir.

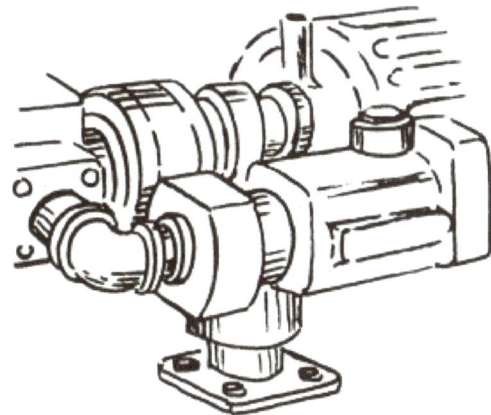

On externally mounted pump inlet strainers, element can be removed without disconnecting filter body from the line.

Figure 13.15- Disassembling Pump Strainer (Courtesy of Womack)

Heavily Varnished Strainer

Damaged Suction Filter

Figure 13.16- Pump Strainer that Should be Replaced

❖ Step 2: Pump and Relief Valve (Refer to Fig. 13.14):

If step 1 didn't solve the problem, do the following:

- Isolate the pump and relief valve from the rest of the system by disconnecting the plumbing (at Point B), and cap both ends of the disconnected lines.
- Fully open the relief valve, run the pump at the right speed across the relief valve.
- Gradually close the relief valve and observe pump pressure (at Point A) and flow through the relief valve (at Point C).
- If the pressure builds-up as the relief valve is tightened, and pump flow through the PRV stays fairly constant (as shown in Fig. 13.17), then neither the pump nor the PRV is at fault. Pump flow may decrease slightly due to normal rated internal leakage.
- Pump is bad if the pump flow through the PRV at high pressure remarkably decreases.
- Pressure relief valve is bad if, while tightening the PRV, full flow is observed through the PRV but the pressure does not rise above a low value, say 10 bar (145 psi).

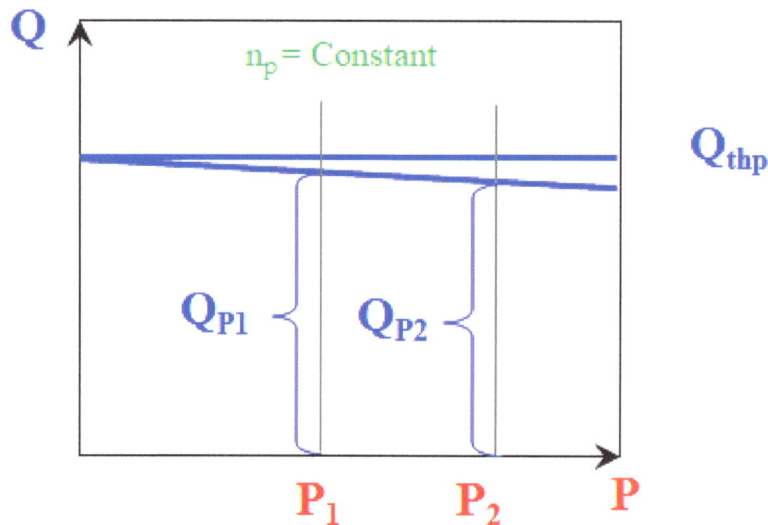

Figure 13.17- Flow-Pressure Characteristics of a Fixed Displacement Pump

❖ Step 3: Pump:

If step 2 shows that the pump is at fault:

- Check for power take-off from the prime mover, slipping belts, sheared shaft key or pin, broken shaft, broken coupling.
- Check for high working temperature increases the internal leakage.
- Disassemble the pump and inspect the internal parts for wear.
- Perform detailed inspection by applying the chart: **"T-Pump-02: Low Flow out of the Pump"**.

❖ **Step 4: Pressure Relief Valve:**

If step 2 shows that the pressure relief valve is at fault, check the following:
- For a quick proof that PRV is at fault, replace it with a new one and repeat the test process.
- Check the valve body and its connection to a subplate or pipes.
- Disassemble the valve, check if there is any broken element or weakened spring.
- Check also for free movement of the spool or poppet.
- Blow air in the orifice of pilot-operated relief valves.
- Clean the valve and assemble it in the system.
- Perform detailed inspection by applying the chart: **"T-Valve-03: PCV Troubleshooting"**.

❖ **Step 5: Cylinder:**

If the tested components so far were found working fine, then inspect the cylinder as follows:
- **External Leakage:** As shown in Fig. 13.18, external leakage around cylinder rod seal or at the cap ends can be easily detected. As shown in the figure, small leakage is observed as drops of oil and accumulated dirt around it. Excessive leakage is observed as a continuous stream of oil leaking out of the cylinder.
- **Internal Leakage:** Leakage across piston seals is not as easy to detect but will usually slow the cylinder and drifts the load. To test for internal leakage, as shown in Fig. 13.19:
  - Fully retract the cylinder and apply full pressure on the rod side.
  - Observe the leakage from the piston side cylinder port.
  - Fully extend the cylinder.
  - Repeat the test to check leakage across the piston from the other side.
- Perform detailed inspection by applying **"T-Cylinder-01: Cylinder Troubleshooting"**.

Small Leakage   Excessive Leakage

Figure 13.18- External Leakage in Hydraulic Cylinders

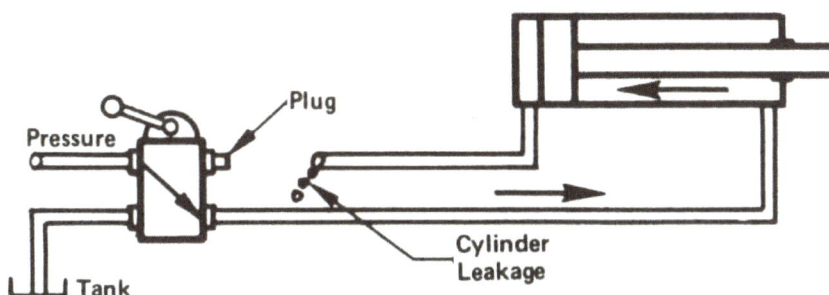

Figure 13.19- Internal Leakage Test of Hydraulic Cylinders (Courtesy of Womack)

❖ **Step 6: Directional Control Valve:**

If the tested components so far were found working fine, then inspect the DCV as follows:

- It is not common that a spool of a directional valve becomes worn so badly that it will leak the entire volume of the pump under pressure. However, spool leakage is a more serious problem when using a pump of small displacement operating at very high pressure.
- For a quick proof that DCV is at fault, replace it with a new one and repeat the test process.
- As shown in the previous figure, spool leakage can be tested as follows:
  - Disconnect the tank return line.
  - Observe leakage flow through the tank port of the valve.
- Perform detailed inspection by applying chart **"T-Valve-01: DCV Troubleshooting"**.

13.11- Troubleshooting of Closed Hydraulic Circuit (Hydrostatic Transmission)

As shown in Fig. 13.20, a closed hydraulic circuit is the one that has the return oil from the motor sent back to the pump. It is also referred to as *"Hydrostatic Transmission"*.

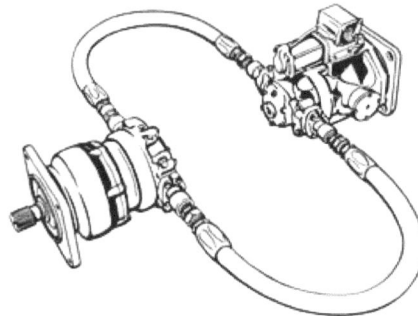

Figure 13.20- Hydrostatic Transmission (Courtesy of Womack)

Before going to troubleshoot a hydrostatic transmission, the following should be prepared:

- A copy of the factory service manual for the machine to be serviced.
- 400-700 bar (6000 – 10000 psi) high-pressure gauge.
- 35 bar (500 psi) low-pressure gauge.
- Port adaptors for connecting the gauges.
- Short lengths of high-pressure and low-pressure hose and fittings.
- Usual mechanics tools.

Typical hydrostatic transmission circuit diagram isn't that simple. Most hydrostatic transmissions contain variable pump driven by an engine. Figure 13.21 shows a typical hydrostatic transmission circuit diagram that, in addition to the main pump and the hydraulic motor, contains the following components:

1- Charge pump.
2- Check valves.
3- Pressure relief valve for the charge pump.
4- Low pressure side hot oil shuttle valve.
5- Hot oil relief valves.
6- Overload pressure relief valves for the two sides of the hydrostatic transmission.

In a closed circuit (hydrostatic transmission), shown in Fig. 13.21, a low side hot oil shuttle valve is used to discharge supercharge oil from the low side of the circuit for cooling. This shuttle valve is piloted from the pump work ports, with the highest pressure shifting the valve so that the opposite (low pressure) side of the loop is connected to tank (typically through a hot oil relief valve, into the motor case). When the pump is commanded to neutral, both sides of the circuit see equal charge pressure, and the shuttle valve centers, causing all charge oil to exit the circuit through the charge pump relief.

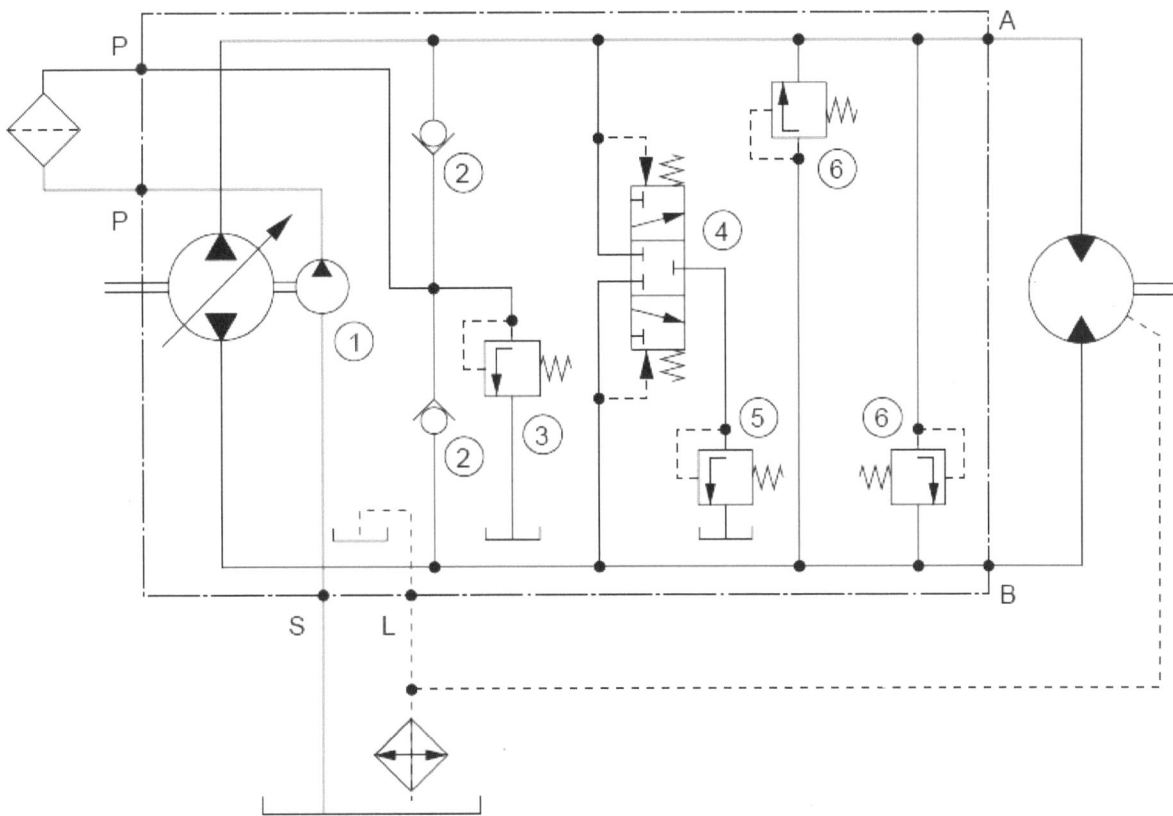

Figure 13.21- Typical Circuit Diagram of Hydrostatic Transmission

Be reminded to avoid the following common mistakes:
- DO NOT screw taper pipe threads into straight thread. Most transmissions have no taper pipe threads on them.
- DO NOT ever plug the case drain.
- DO NOT change the low-pressure relief valve setting on transmissions which were previously running normally.
- DO NOT change the setting of the high-pressure relief valves unless you have the instruments and the factory instruction manual showing how to re-set them. They must be set to about 500 PSI higher than the pressure compensator setting in the manner outlined in the manual.

Table 13.11 shows the relevant troubleshooting chart.

T-System-09-Troubleshooting of Closed Hydraulic Circuits (Hydrostatic Transmission)	
Prechecks:	<ul><li>Whenever a problem occurs, always check first:</li><li>Oil levels.</li><li>Power transmission to the main pump for possible damage, e.g. broken shafts or couplings, slipping belts, etc.</li><li>Control linkages.</li><li>Gaskets and seals.</li><li>Inspect the high-pressure hoses or pipes between the pump and motor and replace any suspected lines.</li></ul>
System is excessively hot?	<ul><li>Consult Chart:</li><li>**"T-System-04-Excessive System Heat".**</li></ul>
System response is sluggish?	<ul><li>Check setting of main pump control.</li><li>Check charge pump low pressure and low-pressure PRV **(See Note 1).**</li></ul>
Transmission operates in one direction only, or problem shows up only in forward or reverse motion?	<ul><li>Check main pump control.</li><li>Check leaking, sticking, and setting of the PRV for the nonworking side of the transmission.</li><li>Check Shuttle Valve.</li></ul>
<ul><li>Loss of power and/or speed in either direction?</li><li>Motor may run when unloaded but will not produce full torque or speed.</li></ul>	<ul><li>Check engine for correct no-load RPM.</li><li>Run engine at load and check for proper performance.</li><li>Check fluid level in the reservoir.</li><li>Check inlet filter.</li><li>Check low pressure of charge pump **(See Note 1).**</li><li>Check high pressure of main pump **(See Note 2).**</li></ul>
Excessive Case Drain **(See Note 3).**	Check case drain considering the following:<ul><li>Check for the pump and the motor separately.</li><li>Combined case drain won't give indication which component is at fault.</li><li>Check case drain when pump displacement increases.</li><li>Check case drain when load is applied to the system.</li><li>Compare measured case drain with the rated values.</li><li>Excessive case drain flow indicates serious problem in the pump.</li></ul>

**Table 13.11- Troubleshooting Chart
(T-System-09-Troubleshooting of Closed Hydraulic Circuits)**

Note 1: Checking Low Pressure of Charge Pump:

❖ Complete loss of charge pump pressure at any percentage of the main pump displacement is due to:

- Broken drive shaft or coupling to the charge pump.
- Spring breakage, damage, or dirt in the low-pressure relief valve.

❖ Fluctuating of charge pump pressure at any percentage of the main pump displacement, is due to:

- Cavitation of the charge pump.
- Low oil level in the reservoir.
- Collapsed suction hose.
- Dirty inlet filter.

❖ Charge pump pressure drops only when pump displacement increases is due to:

- Worn charge pump is worn out and experiences large internal leakage.

Note 2: Checking High Pressure of Main Pump (excerpted from Womack):

- Block hydraulic motor shaft by breaking the machine or blocking drive wheels.
- **High Pressure Obtained:** As the pump displacement is increased to forward or reverse, the pressure should immediately pick up to the value required by the load. If the vehicle brakes are set, the pressure would immediately pick up to the maximum pressure setting by each side's PRV or pump compensator setting.
- **High Pressure Can't Be Developed (excerpted from Womack):** If there is little or no pressure rise in the system, check the following:
  - ○ **High-pressure relief valves** may be stuck or damaged due to contamination.
  - ○ **Charge pump** isn't properly operative (follow instructions in Note 1).
  - ○ **Power transmission to main pump** may possibly is damaged.
  - ○ **One or both check valves**, which connect the charge pump oil into the low-pressure side, may be damaged, stuck, or leaking. Oil from the main pump may be back-flowing into the charge pump circuit and escaping across the low-pressure relief valve. This would cavitate the loop. Symptoms of this fault would be fairly normal operation of the hydraulic motor when unloaded, but inability to build up high torque for heavy loads. A system with this fault might also show signs of overheating. If the motor will not build up torque in either direction, the check valves are not a likely fault unless by coincidence both check valves started leaking at the same rate.
  - ○ **Leaking shuttle valve** from high pressure side to low pressure side. This could occur due to excessively worn spool.
  - ○ **Main pump controller (pressure compensator)** internal parts may be jammed, dirty, or damaged. If parts of the compensator are removed, it must be re-set to its original setting by service manual procedure.
  - ○ **Excessive case drain** flow, particularly if the flow increases suddenly when the control lever is moved out of neutral, usually indicates very serious damage in the pump or the motor.

Note 3: Checking Case Drain:
The most common type of pump is the pressure-compensating, piston-type pump shown in Fig. 13.22. The tolerances between the pistons and barrel are approximately 0.0004 inch. A small amount of oil at the pump outlet port will bypass through these tolerances and flow into the pump case. The oil is then ported back to the reservoir through the case drain line. This case drain flow does no useful work and is therefore converted into heat.

The normal flow rate out of the case drain line is 1 to 3 percent of the maximum pump volume. For example, a 30-gallon-per-minute (GPM) pump should have approximately 0.3 to 0.9 GPM of oil returning to the tank through the case drain. A severe increase in this flow rate will cause the oil temperature to rise considerably.

To check the flow, the line can be ported into a container of a known size and timed. Unless you have verified that the pressure in the hose is near 0 pounds per square inch (PSI), do not hold the line during this test. Instead, secure it to the container. A flow meter can also be permanently installed in the case drain line to monitor the flow rate. This visual check can be made regularly to determine the amount of bypassing. When the oil flow reaches 10 percent of the pump volume, the pump should be changed. Be aware to use a low pressure drop flow meter so the maximum case drain pressure is below the specified value.

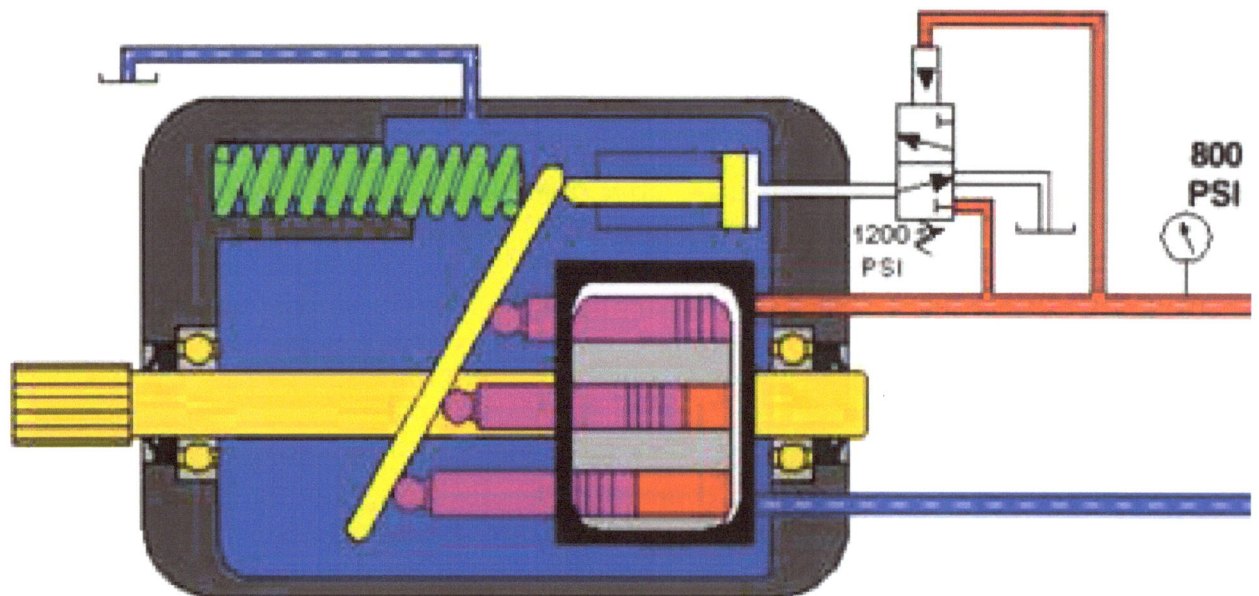

Figure 13.22- Variable-Displacement, Pressure-Compensated Pump (Courtesy of Noria)

13.12- Actuator Slow Performance

The set of troubleshooting charts **T-Actuator-01** through **T-Actuator-07** are applicable for both cylinders and motors. If the fault is not identified by applying these charts, review the following charts if needed: **T-Motors-01 and T-Cylinder-01** in Chapters 5 and 6; respectively.

When a hydraulic actuator performs unusually slow, that results in reduced productivity, increased energy consumption, and automated machines may go out of sequence. Table 13.12 shows the relevant troubleshooting chart. It is to be noted that these charts will be used for both motors and cylinders.

T-System-10-Actuator Slow Performance	
Hydraulic Fluid: Fluid is aerated?	▪ Consult Chart: ▪ **"T-System-01-Fluid Aeration".**
Oil viscosity is too high?	▪ Check oil viscosity and if the machine is too cold.
Pump: Low flow out of the pump?	▪ Consult Chart: **"T-Pump-02-Low Flow out of the Pump".**
Variable displacement pump was set improperly?	▪ Check and reset to the specified value.
Working Temperature: ▪ System is excessively hot?	▪ Consult Chart: ▪ **"T-System-04-Excessive System Heat".**
Accumulator: Accumulator used to boost cylinder speed	▪ Consult Chart: **"T-Accumulator-01-Accumulator Troubleshooting".**
FCV: Flow control valve was set improperly?	▪ Check and reset the valve.
PRV: Pressure relief valve was set too low?	▪ Check and reset the valve.
DCV: Internal leakage in the directional valve?	▪ Check the suspected valve.
Insufficient pilot pressure makes the DCV's spool shifted partially?	▪ Check the pilot pressure of the valve.
Transmission Lines: Restricted pressure lines?	▪ Replace the line.

Table 13.12- Troubleshooting Chart (T-System-10-Atuator Slow Performance)

13.13- Actuator Fast Performance

When a hydraulic actuator performs unusually fast, that results in one or all the following:
- Increased heat generation.
- Automated machines may go out of sequence.

Table 13.13 shows the relevant troubleshooting chart.

T-System-11-Actuator Fast Performance	
Pump: Excessive flow out of the pump?	• Consult Chart: • **"T-Pump-04-Excessive Flow out of the Pump".**
Actuator: Actuator undersized?	• Check and resolve accordingly.
Actuator is a variable displacement motor?	• Check the motor size adjustment. • Check proper motor controller operation.
Actuator speed is automatically controlled?	• Check control system settings. • Check proper speed sensor operation.
Load: Overrunning load?	• Check external load conditions. • Check applied method and relevant valve for controlling overrunning load **(See Note 1).**

Table 13.13- Troubleshooting Chart (T-System-11-Atuator Fast Performance)

Note 1: As shown in Fig. 13.23, an overrunning load can be controlled by a counterbalance valve (1), a throttle-check valve (2), or a pilot-operated check valve. Each method has pros and cons (review volume 1 of this series of textbooks). However, adjustment of the counterbalance valve or the throttle-check valve may change the speed of the load.

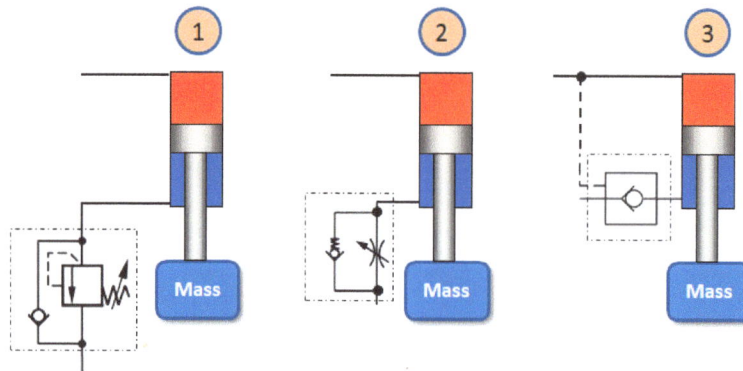

Fig. 13.23- Controlling Overrunning Load

13.14- Actuator Erratic Performance

Actuator *erratic* performance is characterized by a distinct stop-start movement of the actuator.

Erratic motion on an actuator may results in:
- Severe vibration, high pitched noise or chatter.
- Jerky motion and shaking the load that is attached to the actuator.
- As shown in Fig. 13.24, subjecting mobile machines operators to risk of falling.

Table 13.14 shows the relevant troubleshooting chart.

**Fig. 13.24- Actuator Erratic Performance
(Courtesy of Fluid Power Safety Institute)**

T-System-12-Actuator Erratic Performance	
Pump: Erratic flow out of the pump?	▪ Consult Chart: ▪ **"T-Pump-03-Erratic Flow at the Pump Outlet".**
Erratic pressure at the pump outlet?	▪ Consult Chart: ▪ **"T-Pump-07-Erratic Pressure at the Pump Outlet".**
Actuator: Leaking actuator?	▪ Consult Chart: ▪ **"T-System-16-Actuator Leaks".**
Air in actuator is not bled properly?	▪ Bleed air from actuator, review manufacturer's instructions.
Cylinder Stick-Slip motion	▪ Check piston and rod seals condition **(Note 1)**. ▪ Check if there is radial force acting on the rod.
Motor spinning below the rated speed?	▪ Review the motor working speed versus the minimum allowable speed. ▪ Adjust the motor speed accordingly.
Load: Increased friction of outside load?	▪ Check excessive side loading.
Lack of lubrication/greasing to the machine joints and linkages?	▪ Lubricate/grease the joints and linkages.
DCV: Pressure spikes when switching a directional valve?	▪ Unnecessary long hoses used. ▪ Valve switches too fast **(Notes 2)**.
Actuator movement based on shifting a directional valve or an EH valve?	▪ Consult Chart: ▪ **"T-Valve-01- DCV Troubleshooting".** ▪ **"T-Valve-04- EH Valve Troubleshooting".**
System Design: Actuator movement based on a pump connected in parallel to an accumulator?	Check if the pump is loaded/unloaded too often because **(Notes 3)**. ▪ Improper sizing of pump-accumulator combination. ▪ Improper setting of accumulator charge valve or pressure switch.

Table13.14- Troubleshooting Chart (T-System-12-Atuator Erratic Performance)

Note 1 (Stick-Slip): The *Stick-Slip* phenomenon is described as "the spontaneous *jerking* motion that can occur while two objects are sliding over each other." In the area of seals and cylinders, as shown in Fig. 13.25, seals are often thought to be the source of the stick-slip.

Fig. 13.25- Deteriorated Seal causes Stick-Slip Cylinder Motion (Courtesy of Parker)

Note 2 (Pressure Spikes due to Switching DCV): Figure 13.26 shows the construction of a 4/3, closed center, solenoid-actuated, spring-centered, pilot-operated, direct-controlled DCV. The construction and operation of a pilot-operated DCV has been explained in Volume 1 of this series of Textbooks. The figure shows also that the valve is equipped by a *Switching Time Adjustor*. A pilot-operated DCV drives a large flow and drives a large load. When the pilot stage is energized, the main spool is switched very fast and the main pressure port is connected suddenly to the load port so that the load moves in a jerky way. To ease the connection with the load, two meter out throttle-check valves are used to restrict the vented port of the main spool so that the time it takes to switch between different positions is tuned in the field.

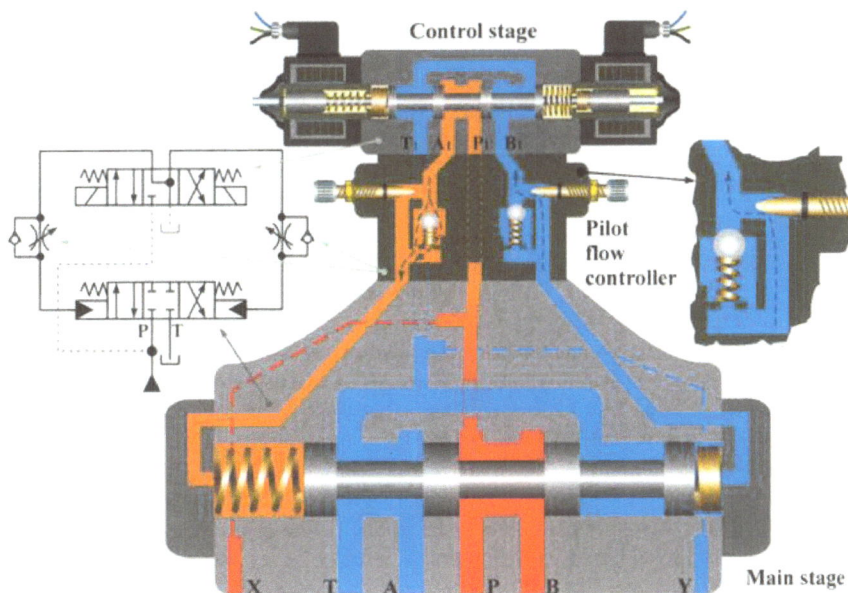

Fig. 13.26 - Pilot-Operated 4/3 Spool-Type Solenoid-Actuated DCV (Courtesy of ASSOFLUID)

Notes 3 (pump is loaded/unladed very often): Figure 13.27 shows that an accumulator (1) is used as an energy storage element. A fixed displacement pump (2) is unloaded through a 2-way, 2-position, normally-open DCV (3). Once the machine is turned ON, a normally-closed pressure switch (4) energizes the DCV (3) so that the pump is loaded. During the dwell time of the machine, the pump (2) has nowhere to go but to fill the accumulator (1). The charging time of the accumulator (1) depends on its size and the size of the pump (2). The pump (2) keeps charging the accumulator (1) until the maximum system pressure is reached. At maximum system pressure, the pressure switch (4) is activated so that the DCV (3) is de-energized and the pump (2) is unloaded. By shifting the main DCV (5), the cylinder (6) consumes a portion of the energy stored at the accumulator (1) and the accumulator pressure reduces accordingly. The accumulator rate of discharge is controlled by a flow regulator (7). When the accumulator pressure reduces, the pressure switch (4) is de-activated; DCV (3) loads the pump to refill the accumulator. The DCV (8) is used to discharge the accumulator automatically as a fail-safe design concept. If the sizes of the pump and accumulator are not compromised, or if the pressure threshold of the pressure switch is narrow, then the pump will be loaded and unloaded more frequently, and the cylinder could move erratically.

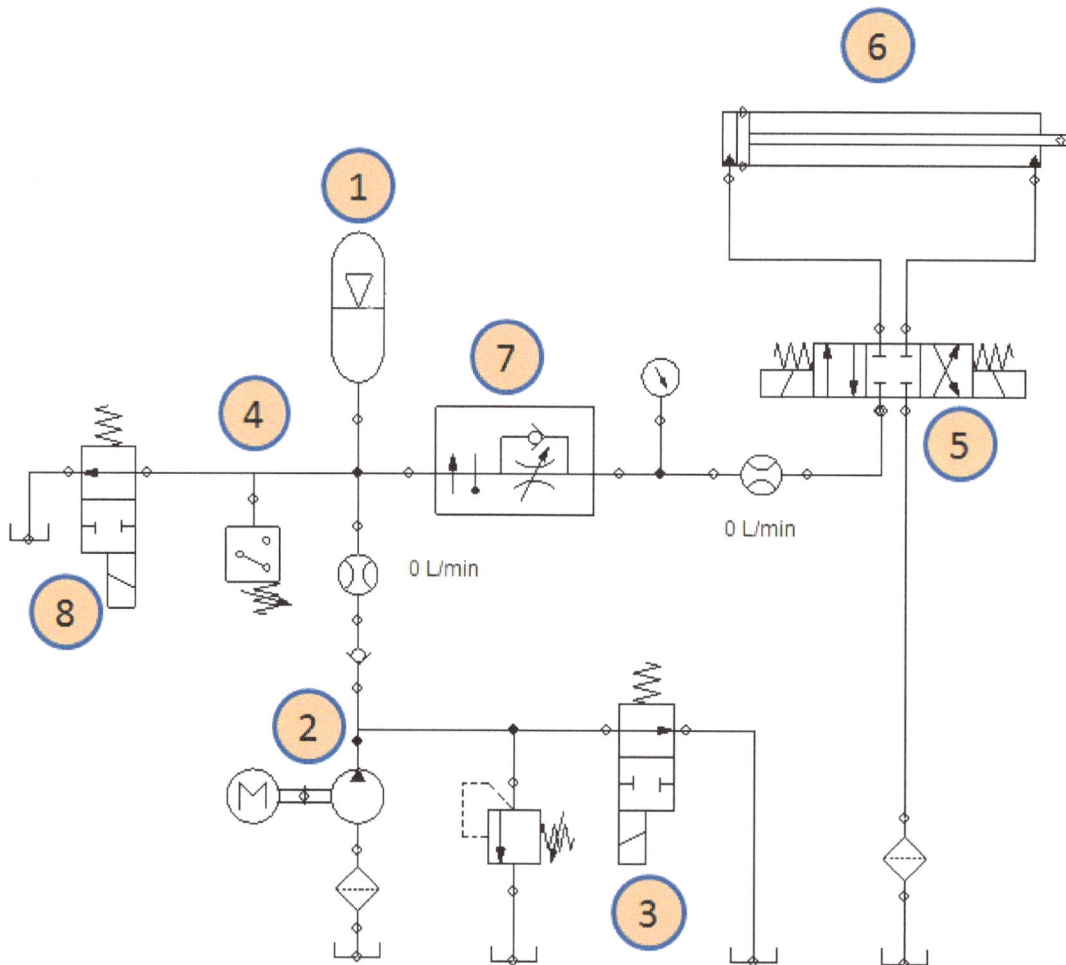

Fig. 13.27-Energy Storage Application Example

13.15- Actuator Moves in Wrong Direction

When a hydraulic actuator moves in the wrong direction, one or all the following can occur:
- Automated machines may go out of sequence.
- Possible machine destruction.
- Risk of personal injury due to unexpected machine movements.

Table 13.15 shows the relevant troubleshooting chart.

T-System-13-Actuator Moves in Wrong Direction	
Pipework: Wrong piping connections with the directional valve or the actuator?	▪ Check and resolve accordingly.
DCV: Directional Valve replaced recently?	Check: ▪ Ordering code. ▪ Proper assembly of valve spool.
Wrong wiring of solenoid-operated directional valve?	▪ Check and resolve accordingly.
Faulty operation of a directional valve	▪ Consult Chart: ▪ **"T-Valve-01-DCV Troubleshooting".**
Control System: Wrong signal generation from the control system?	▪ Check and resolve accordingly.

Table 13.15- Troubleshooting Chart (T-System-13-Atuator Moves in Wrong Directions)

13.16- Actuator Stops to Move

In some cases, an actuator fails to move even when a driving valve is shifted. Table 13.16 shows the relevant troubleshooting chart.

T-System-14-Actuator Stops to Move	
Pump: No flow out of the pump?	▪ Consult Chart: ▪ **"T-Pump-01-No Flow out of the Pump"**.
No pressure at the pump outlet?	▪ Consult Chart: ▪ **"T-Pump-05-No Pressure at the Pump Outlet"**.
System Design: Free recirculation of oil to reservoir being allowed through system? **(See Note1).**	▪ Review the sequence of the operation. ▪ Check the proper operation of open-center and tandem-center valves.
Load: Load is mechanically braked or blocked?	▪ Check braking system if found. ▪ Uncouple the actuator from load and check the operation of the actuator.
Actuator: Actuator moves when the load is reduced?	▪ Consult Chart: ▪ **"T-System-08-Low Power System"**.
Actuator undersized and relief valve open?	▪ Review the size of the actuator or maximum pressure setting.
Actuator broken?	▪ Repair if possible or replace.
Actuator moves based on operation of a sequence valve	▪ Check if the sequence valve was set too high.
Actuator moves based on operation of a check valve	▪ Check if the check valve is jammed. ▪ Check if the pilot operated check valve is jammed or not receiving pilot pressure.
Actuator moves based on shifting a directional or an EH valve?	▪ Consult Chart: ▪ **"T-Valve-01- DCV Troubleshooting"**. ▪ **"T-Valve-04- EH Valve Troubleshooting"**.

Table 13.16- Troubleshooting Chart (T-System-14-Atuator Stops to Move)

Note (1): If a pump is used to drive multiple actuators in parallel, a combination of tandem and/or open center directional control valves can't be used. That is because, as shown in Fig. 13.28, the oil is still able to flow to the tank through the central position of the other directional control valve (DCV).

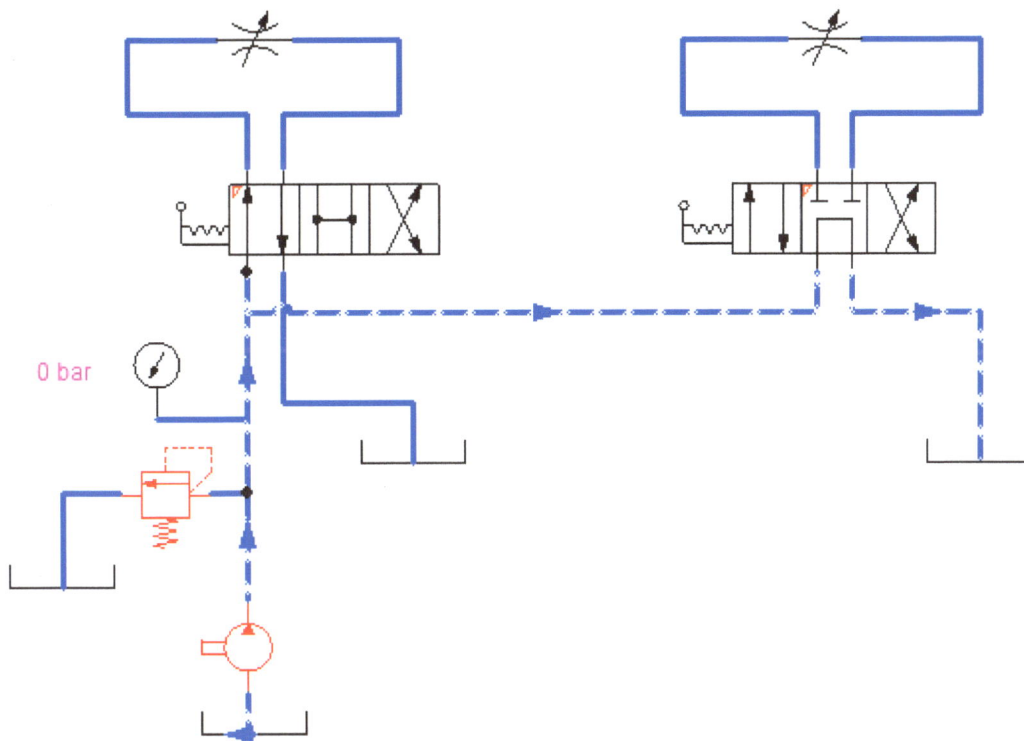

Fig. 13.28- Improper Combination of Tandem and Open Center Valves in Parallel

13.17- Actuator Load Drifts

As shown in Fig. 13.29, handling heavy goods presents particular challenges in terms of technology and demands the highest level of precision and safety. One common fault in such applications is load drifting. Table 13.17 shows the relevant troubleshooting chart.

Fig. 13.29- Heavy Lift Crane (Courtesy from Liebherr)

T-System-15-Actuator Load Drifts	
Actuator: Leaking actuator?	▪ Consult Chart: ▪ **"T-System-16-Actuator Leaks".**
Load: Failure of external mechanical braking system?	▪ Check and act accordingly.
DCV: Null position of directional valve is biased OR directional valve leaks at null?	▪ Consult Chart: ▪ **"T-Valve-01-DCV Troubleshooting".**

Table 13.17- Troubleshooting Chart (T-System-15-Atuator Load Drifts)

13.18- Actuator Leaks

Hydraulic an actuator experience external leakage, one or all the following may occur:
- Missing expensive oil.
- Environmental pollution and increasing cost of removing it.
- Possible actuator loss of power and load failure.

Table 13.18 shows the relevant troubleshooting chart.

T-System-16-Actuator Leaks	
Cylinder rod scored, scratched or bent?	▪ Check and replace rod if needed.
Improper torqueing of tie rods cylinder or motor housing?	▪ Re-torque per manufacturer specifications.
Motor's shaft or cylinder rod subject to radial forces?	▪ Check maximum allowable radial force. ▪ Review the actuator-load attachment.
Piston seal failure (see **Note 1**)?	▪ Consult Chart: ▪ **"T-Seal-01-Seal Troubleshooting"**.

Table 13.18- Troubleshooting Chart (T-System-16-Atuator Leaks)

Note 1 (Refer to Fig. 13.30): If the piston seals are not leak-tight, the cylinder may drift under external force.

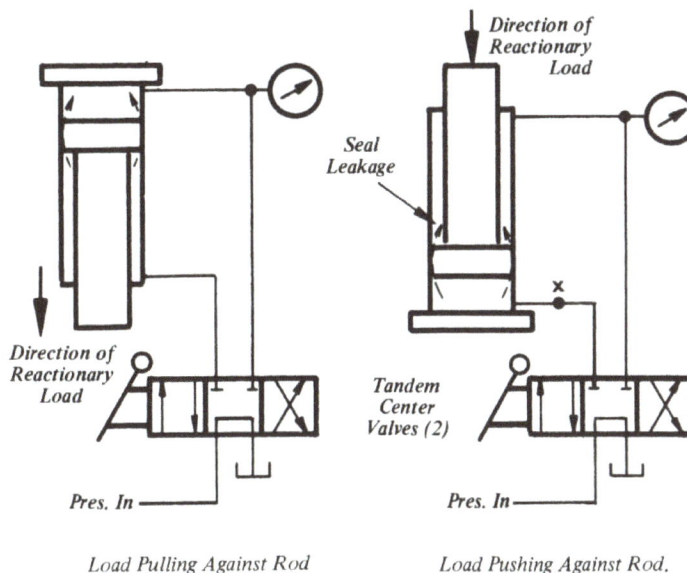

Fig. 13.30- Cylinder Drift due to Piston Seal Leaking (Courtesy from Womack)

Chapter 14
Examples of Hydraulic Systems Troubleshooting

Objectives

In this chapter several case studies are presented as examples of applying the logic trouble shooting methodology for hydraulic systems fault detection. In addition, troubleshooting case studies following analytical fault detection methodology are presented. Examples were chosen from both industrial and mobile applications.

Brief Contents

14.1-Case Studies Using Logic Fault Detection Methodology

14.2-Industrial Applications Case Studies Using Analytical Fault Detection Methodology

14.3-Mobile Applications Case Studies Using Analytical Fault Detection Methodology

Chapter 14: Examples of Hydraulic Systems Troubleshooting

14.1-Case Studies Using Logic Fault Detection Methodology

As it has been discussed in detail in Chapter 1, the following ten steps form the heart of the logic fault detection procedure:

- ❖ **Step 1: Review Safety Instructions:**
- ❖ **Step 2: Review Machine History:**
- ❖ **Step 3: Identify Main System Fault:**
- ❖ **Step 4: Apply the System-Level Troubleshooting Chart:**
- ❖ **Step 5: List Suspicious Components:**
- ❖ **Step 6: Perform Preliminary Check on Suspicious Components:**
- ❖ **Step 7: Apply Detailed Check on Suspicious Components:**
- ❖ **Step 8: Fault Evaluation Decision for Repair or Replacement:**
- ❖ **Step 9: Startup and Testing:**
- ❖ **Step 10: Future Considerations and Documentation:**

The following examples present practical applications of this methodology for troubleshooting hydraulic systems. Examples were selected from common machines in industrial and mobile applications.

14.1.1- Slow Actuator on a Manufacturing Machine

Figure 14.1 shows a hydraulic driven *manufacturing machine*. A slow actuator was reported by the machine operator.

Fig. 14.1 – Hydraulic Driven Manufacturing Machine

❖ **Step 1: Review Safety Instructions:**
  ▪ Safety instructions of the machine/work environment were reviewed.

❖ **Step 2: Review Machine History:**
  ▪ By reviewing the machine history, it was found to be working fine before unexpectedly the actuator started slowing down and machine productivity is reduced.

❖ **Step 3: Identify Main System Fault:**
  ▪ Slow Actuator Performance.

❖ **Step 4: Apply the System-Level Troubleshooting Chart:**
  ▪ "T-System-10: Actuator Slow Performance"
  ▪ Aeration in system is identified →
  ▪ "T-System-01-Fluid Aeration Fluid Aeration" →
  ▪ "T-Pump-11-Air Leaks into the Pump" →
  ▪ Leaking fitting on suction line was identified.

❖ **Step 5: List Suspicious Components:**
  ▪ SKIP.

❖ **Step 6: Perform Preliminary Check of the Suspicious Components:**
  ▪ Skip.

❖ **Step 7: Apply Detailed Check of the Suspicious Components:**
  ▪ Skip.

❖ **Step 8: Fault Evaluation Decision for Repair or Replacement:**
  ▪ Tighten the leaking fitting on intake line.

❖ **Step 9: Startup and Testing:**
  ▪ Completed

❖ **Step 10: Future Considerations and Documentation:**
  ▪ Discuss and resolve reasons why fittings on intake line become loose over the time.
  ▪ Add the step of checking this fitting among the routine maintenance schedule.

14.1.2- Interrupted Duty Cycle of a Manufacturing Machine

A hydraulic driven *manufacturing machine* performs a certain duty cycle, in part of which a hydraulic cylinder is reciprocated few times. The machine works fine for few days before the duty cycle was interrupted and so that the operator trips the machine and ask for a repair.

❖ **Step 1: Review Safety Instructions:**
  ▪ Safety instructions for the machine/work environment were reviewed.

❖ **Step 2: Review Machine History:**
  ▪ By reviewing the machine history and the circuit diagram (Fig. 14.2), it was found that a PLC is used to generate the required sequence of signals to meet the requirement of the duty cycle. The machine was working fine before the duty cycle was recently modified to increase the machine productivity. No parts were replaced.

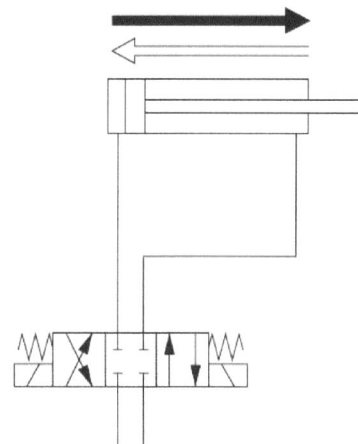

Fig 14.2 – Hydraulic Circuit Diagram of the Investigated Case

❖ **Step 3: Identify Main System Fault:**
  ▪ Actuator Stop to Move.

❖ **Step 4: Apply the System-Level Troubleshooting Chart:**
  ▪ "T-System-14: Actuator Stops to Move" →
  ▪ None of the listed reasons were identified.

❖ **Step 5: List Suspicious Components:**
  ▪ DCV that reciprocates the cylinder.

❖ **Step 6: Perform Preliminary Check of the Suspicious Components:**
  ▪ DCV valve was inspected using the relevant inspection sheet (Table 14.1).
  ▪ EH Directional Control valve Data Sheet was reviewed (Fig. 14.3).
  ▪ Start with "T-Unit-01-General Check" → No fault was identified

❖ **Step 7: Apply Detailed Check of the Suspicious Components:**
  ▪ "T-Valve-04- EH Valve Troubleshooting" → Valve spool isn't moving → solenoid is burn out.

❖ **Step 8: Fault Evaluation Decision of Repair or Replacement:**
- Burn out solenoid is replaced. Same problem occurred after few days.
- It was noted that this valve has a switching rate of 7200/hr = 2 Hz. The modified control code was found to increase the duty cycle frequency to 4 Hz.
- Solution: Replace the current valve by an equivalent DC-Driven EH-valve that has higher switching frequency.

❖ **Step 9: Startup and Testing:**
- Completed

❖ **Step 10: Future Considerations and Documentation:**
- Advice control personnel to review specs of hydraulic components before changing control codes. Basic and EH training is crucial for control personnel.

Hydraulic Valve Inspection Sheet	
Manufacturer	Bosch Rexroth
Model #	4WE 6 E6X
Serial #	NA
Location	NA
Pressure Control Valve Type	▪ Direct [☐Relief ☐Counterbalance ☐Sequence ☐Reducing] ▪ Pilot [☐Unloading ☐Over-Center ☐ Motor Brake]
Directional Control Valve Type	▪ # Ports (4) # Positions (3) ▪ Initial/Central Position: (Closed-Center) ▪ Reset [☐Spring ☐Detent] ▪ Actuation: [☐Manual ☐Mechanical ☐Pilot ☐Electrical] ▪ More info ( )
Flow Control Valve Type	☐Throttle ☐Regulator
EH Valve	Type: [☐ON/OFF ☐Proportional ☐Servo] Signal: ( ) Current = Voltage = 110 AC Power:
Valve Configuration	Operation: [☐Direct "Single-Stage" ☐ Pilot "Multiple stages"] Control: [☐Direct "Internal" ☐ Pilot "External"] Drain: [☐ Internal ☐ External] Built-in Check Valve [☐ Yes ☐ No]
Moving Element:	☐ Poppet Type ☐ Spool Type [☐Linear ☐Rotary]
Mounting	☐ Subplate ☐ Line ☐ Manifold "Screw-In" ☐ Sandwich ☐ Other:
Ports/Flow	Port size = Rated flow Rate =
Conditions	Parts: Seals:
Other Notes:	

Table 14.1 – Hydraulic Valves Inspection Sheet

1. Valve body
2. Two Solenoids
3. Spool
4. Two Centering Springs
5. Plunger
6. Plastic Cover
7. Optional Manual Override

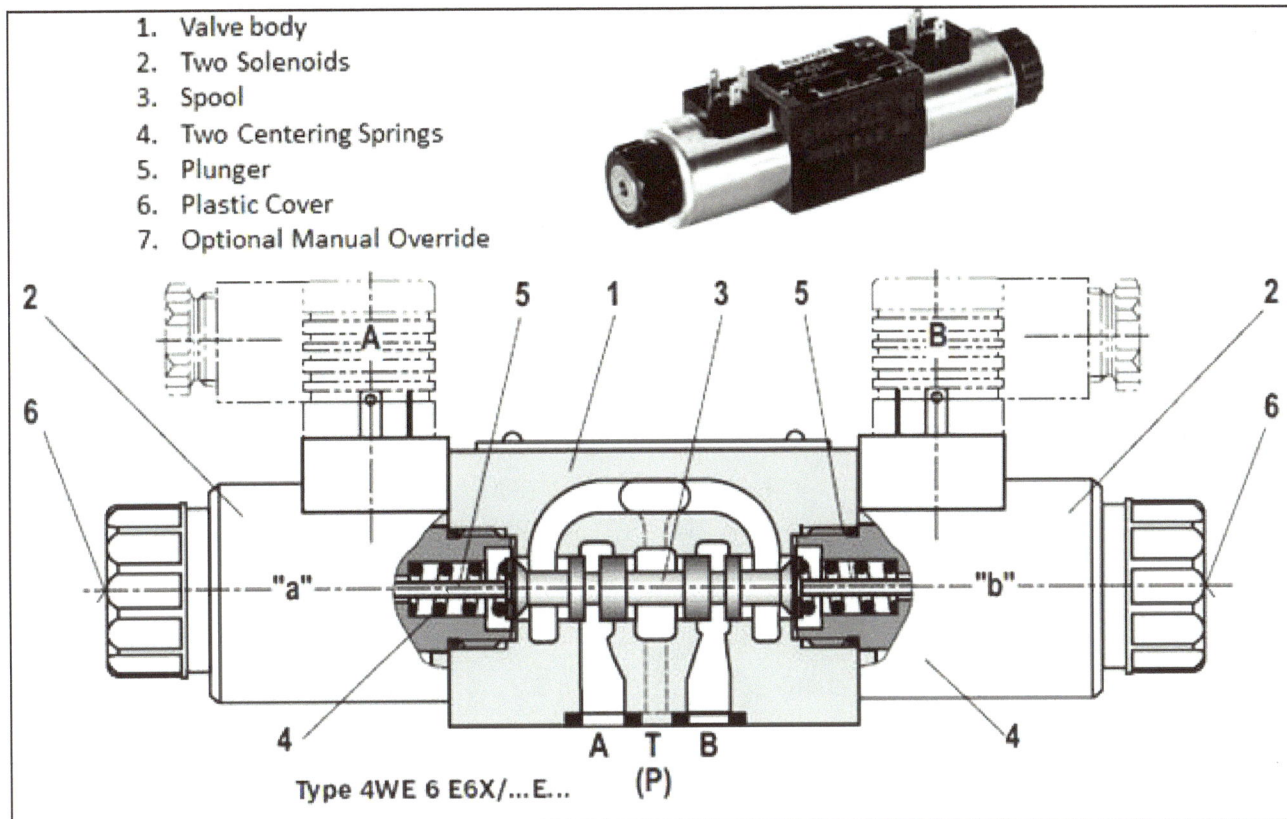

Type 4WE 6 E6X/...E...

Technical data

electric

Voltage type		Direct voltage	Alternating voltage 50/60 Hz
Available voltages	V	12, 24, 96, 205	110, 230
Voltage tolerance (nominal voltage)	%	±10	
Power consumption	W	30	–
Holding power	VA	–	50
Switch-on power	VA	–	220
Duty cycle	%	100	
Switching time according to ISO 6403 – ON	ms	25 ... 45	10 ... 20
– OFF	ms	10 ... 25	15 ... 40
Maximum switching frequency	1/h	15000	7200
Maximum surface temperature of the coil [4]	°C [°F]	120 [248]	180 [356]

Fig 14.3 – Data Sheet of Inspected EH Directional Valve

14.1.3- A Winch Failed to Move

A hydraulic driven *winch* is operated remotely using a pilot operated directional EH-valve. The winch works fine before it stops to move.

❖ **Step 1: Review Safety Instructions:**
- Safety instructions for the machine/work environment were reviewed.

❖ **Step 2: Review Machine History:**
- By reviewing the machine history and the circuit diagram (Fig. 14.4), it was found that the spool of the Main stage was heavily worn and was replaced by another one.

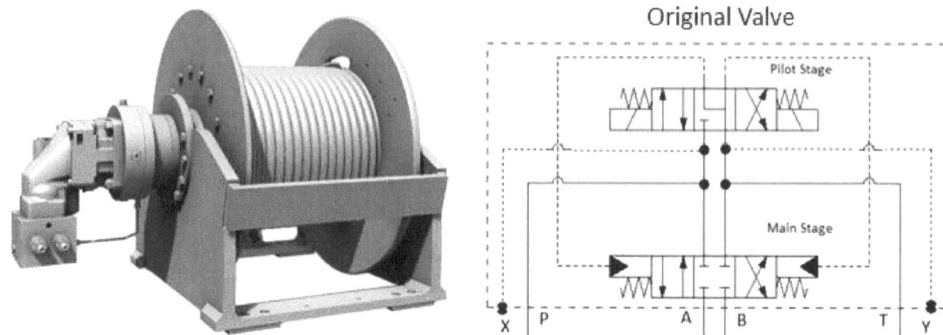

Fig 14.4 – Hydraulic Circuit Diagram of the Investigated Case

❖ **Step 3: Identify Main System Fault:**
- Actuator Stop to Move.

❖ **Step 4: Apply the System-Level Troubleshooting Chart:**
- "T-System-14: Actuator Stops to Move".→
- This chart was investigated, none of the listed reasons where identified except that this machine was driven by an EH directional Valve.

❖ **Step 5: List Suspicious Components:**
- DCV that operates the winch.

❖ **Step 6: Perform Preliminary Check of the Suspicious Components:**
- DCV valve was inspected using the relevant inspection sheet (Table 14.2).
- EH Directional Control valve Data Sheet was reviewed.
- Start with "T-Unit-01-General Check" → model number was found different → Main spool wasn't selected properly. It is noticed that the original valve of a closed-center type is replaced by other spool of an open-center type.

❖ **Step 7: Apply Detailed Check of the Suspicious Components:**
- Skip

Hydraulic Valve Inspection Sheet	
Manufacturer	Bosch Rexroth
Model #	NA
Serial #	NA
Location	NA
Pressure Control Valve Type	• Direct [☐Relief ☐Counterbalance ☐Sequence ☐Reducing] • Pilot [☐Unloading ☐Over-Center ☐ Motor Brake]
Directional Control Valve Type	• # Ports (4) # Positions (3) • Initial/Central Position: (Closed-Center) • Reset [☐Spring ☐Detent] • Actuation: [☐Manual ☐Mechanical ☐Pilot ☐Electrical] • More info ( )
Flow Control Valve Type	☐Throttle ☐Regulator
EH Valve	Type: [☐ON/OFF ☐Proportional ☐Servo] Signal: ( ) Current = Voltage = 110 AC Power:
Valve Configuration	Operation: [☐Direct "Single-Stage" ☐ Pilot "2-stages"] Control: [☐Direct "Internal" ☐ Pilot "External"] Drain: [☐ Internal ☐ External] Built-in Check Valve [☐ Yes ☐ No]
Moving Element:	☐ Poppet Type ☐ Spool Type [☐Linear ☐Rotary]
Mounting	☐ Subplate ☐ Line ☐ Manifold "Screw-In" ☐ Sandwich ☐ Other:
Ports/Flow	Port size = Rated flow Rate =
Conditions	Parts: Seals:
Other Notes:	Spool of the main stage is open-center type.

Table 14.2 – Hydraulic Valves Inspection

❖ **Step 8: Fault Evaluation Decision of Repair or Replacement:**
- If an open-center spool is used in the main stage of a pilot-operated directional valve, one of the following solutions can resolve the problem.
- Solution 1: Use external (pilot) source to supply the control pressure for the pilot stage.
- Solution 2: As shown in Fig. 14.5, place a spring-loaded check valve on the pressure port of the main stage.

❖ **Step 9: Startup and Testing:**
- Completed.

❖ **Step 10: Future Considerations and Documentation:**
  ▪ Circuit diagram must be updated as shown in Fig. 14.5.

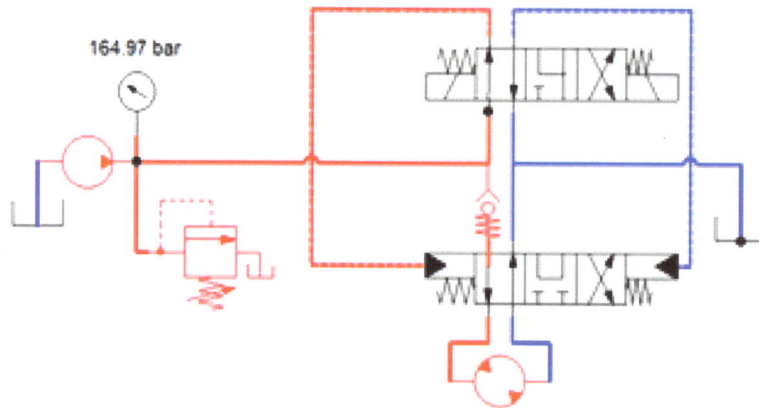

164.97 bar

Fig 14.5 – Updated Hydraulic Circuit Diagram to Resolve the Problem

14.1.4- EH Cylinder Deceleration System Isn't Working Properly

An EH *cylinder-deceleration* system, shown in Fig. 14.6, isn't operating properly. The cylinder supposed to decelerate when the cylinder approaches the position S2. The system was working fine before it was noticed that the cylinder extends with the same speed after passing position S2.

❖ **Step 1: Review Safety Instructions:**
  ▪ Safety instructions for the machine/work environment were reviewed.

❖ **Step 2: Review Machine History:**
  ▪ Reviewing the machine history and circuit diagram →
  ▪ No recent changes were reported.

❖ **Step 3: Identify Main System Fault:**
  ▪ Faulty system sequence.

❖ **Step 4: Apply the System-Level Troubleshooting Chart:**
  ▪ "T-System-06: Faulty System Sequence" → → power line of the switch S2 was accidentally disassembled from the mounting terminal.

❖ **Step 5: List Suspicious Components:**
  ▪ Skip.

❖ **Step 6: Perform Preliminary Check of the Suspicious Components:**
  ▪ Skip

❖ **Step 7: Apply Detailed Check of the Suspicious Components:**
  ▪ Skip

❖ **Step 8: Fault Evaluation Decision for Repair or Replacement:**
  ▪ Switch S2 is wired properly.
  ▪ Main chassis, on which the control panel is mounted, is subjected to permanent vibration. Control panel is relocated to another fixed frame.

❖ **Step 9: Startup and Testing:**
  ▪ Completed

❖ **Step 10: Future Considerations and Documentation:**
  ▪ Control panels should be located apart from the machine body if possible. Otherwise, it must be located apart from vibrating chasses.

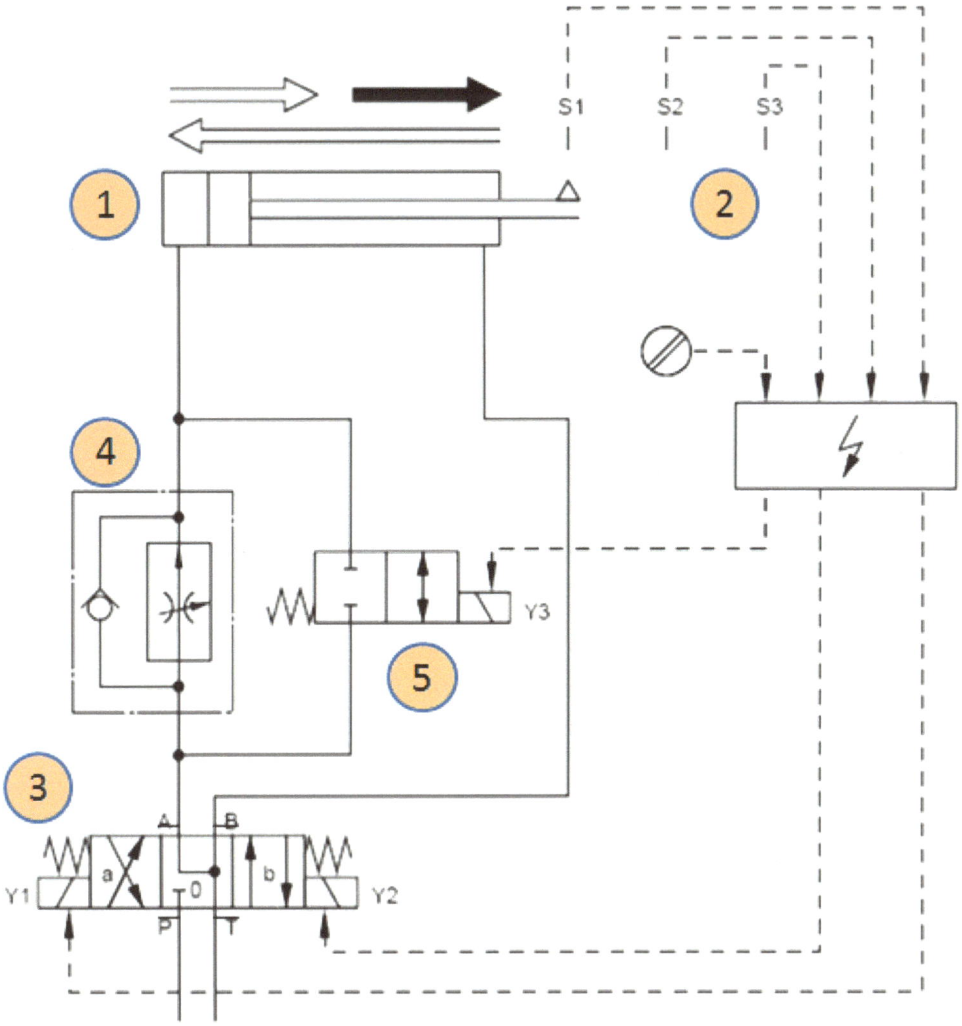

Fig. 14.6- Electro-Hydraulic Cylinder Deceleration Circuit (Courtesy of Bosch Rexroth)

14.1.5- Steady State Error in Cylinder Position Control System

An EH motor in a *speed control* system, shown in Fig. 14.7, is experiencing steady state error. The motor runs slower than the desired reference speed set by the control system.

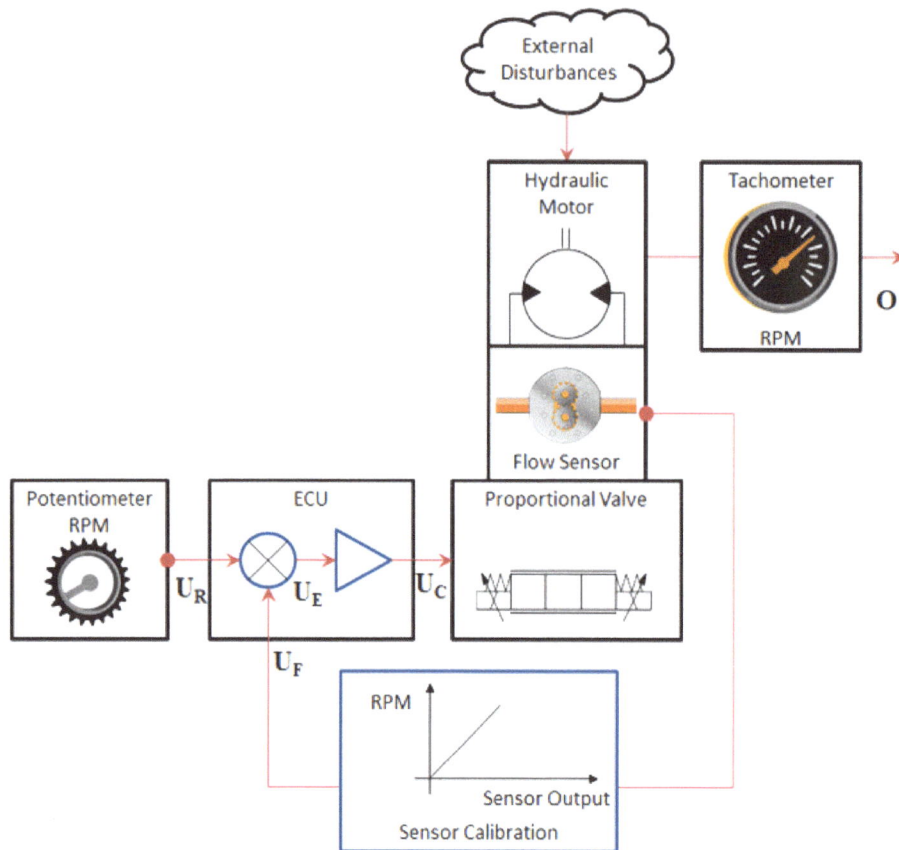

Fig. 14.7- Electro-Hydraulic Cylinder Position Control System

❖ **Step 1: Review Safety Instructions:**
  ▪ Safety instructions for the machine/work environment were reviewed.

❖ **Step 2: Review Machine History:**
  ▪ By reviewing the machine history, it was found that the machine was working fine before it required some calibration to remove a steady state error that resulted over the time. Even after this calibration process, steady state error starts to appear again after a month of operation.
  ▪ Hydraulic and control circuit diagrams were also reviewed. It was noticed that the control loop is closed on the flow rate that drives the motor.

❖ **Step 3: Identify Main System Fault:**
- Actuator slow performance.

❖ **Step 4: Apply the System-Level Troubleshooting Chart:**
- "T-System-10: Actuator Slow Performance" → No fault was identified.

❖ **Step 5: List Suspicious Components:**
- Any elements in the system could be suspicious. However, since the control loop is closed on the flow, this means that the control system is not able to detect errors resulting from motor. Then, wisely, a decision has been made to check the motor first before spending time checking other components.

❖ **Step 6: Perform Preliminary Check of the Suspicious Components:**
- Motor was inspected using the relevant inspection sheet.
- Motor Data Sheet was reviewed.
- "T-Unit-01-General Check" → No fault was identified.

❖ **Step 7: Apply Detailed Check of the Suspicious Components:**
- "T-Motor-01-Motor Troubleshooting" → "T-Pump-10-Excessive Pump Wear (As applied for Motors)" → High water content in the fluid.

❖ **Step 8: Fault Evaluation Decision for Repair or Replacement:**
- Increasing water content in the fluid caused gradual increase in wear of the internal rotating elements. As a result, gradual increase of internal leakage. So, the motor was gradually slowing down without being detected by the control system. Motor was replaced.
- Since the system is small, it was advised to drain, flush, and change the fluid.
- Source of water penetration into the system was investigated and resolved.

❖ **Step 9: Startup and Testing:**
- Completed.

❖ **Step 10: Future Considerations and Documentation:**
- Frequent fluid analysis is required.

14.1.6- Loss of Power Accompanied by an Increase of Pump Noise

In the general hydraulic system, shown in Fig. 14.8, a gradual or sudden loss of high pressure, resulting in loss of power or speed of the cylinder. It was also noticed that the cylinder may stall under light loads or may not move at all. The loss of power was accompanied by an increase in pump noise, especially as the pump tries to build up pressure.

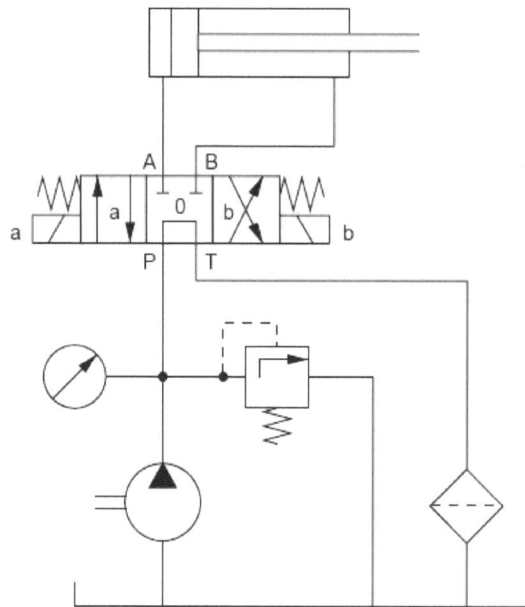

Fig. 14.8- Hydraulic Circuit of the Investigated Case

❖ **Step 1: Review Safety Instructions:**
  ▪ Safety instructions for the machine/work environment were reviewed.

❖ **Step 2: Review Machine History:**
  ▪ Reviewing the machine history didn't provide any reasons for the indicated symptoms, except that the hydraulic fluid was recently replaced.

❖ **Step 3: Identify Main System Fault:**
  ▪ Pump noise and loss of power when building pressure.

❖ **Step 4: Apply the System-Level Troubleshooting Chart:**
  ▪ "T-Pump-12: Excessive Pump Noise & Vibration" →
  ▪ "T-System-02: Pump Cavitation" → Fluid is too viscous.

❖ **Step 5: List Suspicious Components:**
  ▪ Skip.

❖ **Step 6: Perform Preliminary Check of the Suspicious Components:**
- Skip.

❖ **Step 7: Apply Detailed Check of the Suspicious Components:**
- Skip.

❖ **Step 8: Fault Evaluation Decision for Repair or Replacement:**
- Replacing the hydraulic fluid with another of higher viscosity results in creating highly negative pressure in the suction line. Fluid was drained and replaced with fluid specified by the machine manufacturer.

❖ **Step 9: Startup and Testing:**
- Completed.

❖ **Step 10: Future Considerations and Documentation:**
- Adhere to specifications published by the machine manufacturer.

14.1.7- Mold of an Injection Molding Machine is Partially Filled

In a hydraulic-driven *injection molding* machine, shown in Fig. 14.9, over the time it was noticed that the mold partially filled resulting in batches inconsistency.

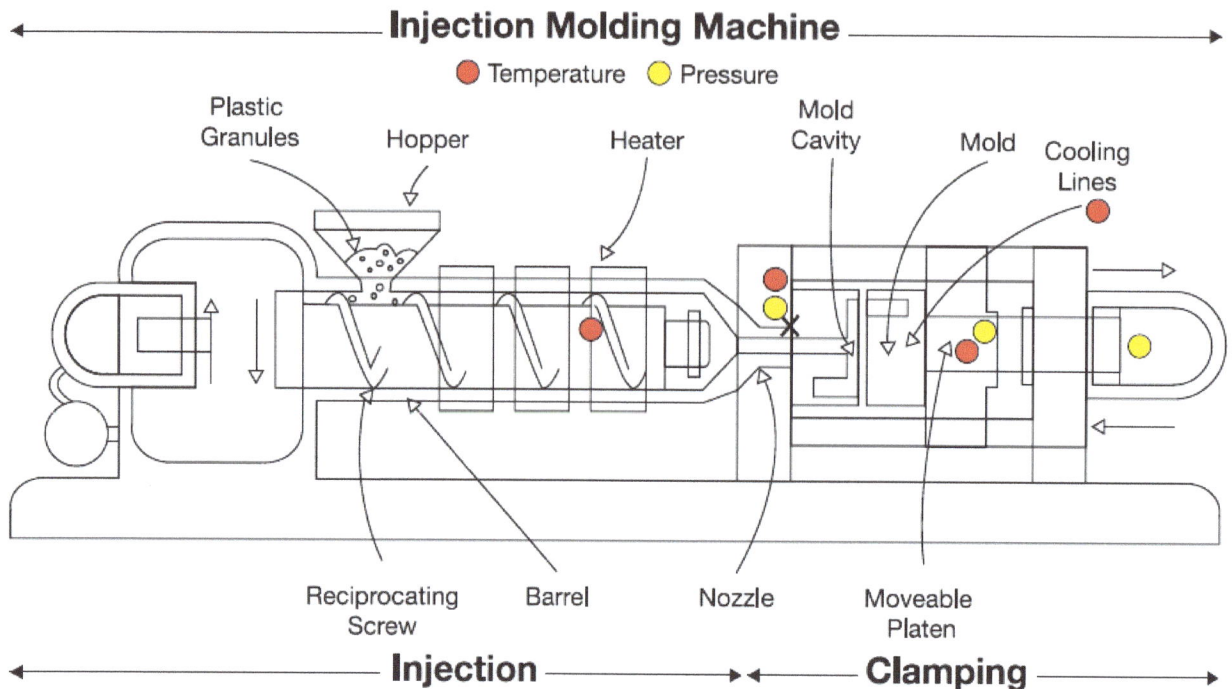

Fig. 14.9- Hydraulic-Driven Injection Molding Machine (Courtesy of Parker)

❖ **Step 1: Review Safety Instructions:**
- Safety instructions for the machine/work environment were reviewed.

❖ **Step 2: Review Machine History:**
- Reviewing the machine history didn't provide any reasons for the indicated symptoms.
- Service manual of the machine is reviewed.

❖ **Step 3: Identify Main System Fault:**
The main reason why a mold isn't fully filled is loss of injection pressure, that can be considered as low power fault.

❖ **Step 4: Apply the System-Level Troubleshooting Chart:**
- "T-System-5: Low Power System" → "T-Pump-06: Low Pressure at the Pump Outlet".
- → Pressure relief is set too low.

❖ **Step 5: List Suspicious Components:**
  ▪ Skip.

❖ **Step 6: Perform Preliminary Check of the Suspicious Components:**
  ▪ Skip.

❖ **Step 7: Apply Detailed Check of the Suspicious Components:**
  ▪ Skip.

❖ **Step 8: Fault Evaluation Decision for Repair or Replacement:**
  ▪ Reset the relief valve.

❖ **Step 9: Startup and Testing:**
  ▪ Completed.

❖ **Step 10: Future Considerations and Documentation:**
  ▪ Add a wireless sensor, shown in Fig. 14.10, that can detect the slightest change in pressure and report that information to a user's mobile device, which will alert users to any pressure drop allowing them to diagnose fault immediately.

Pressure Sensor Features
- For commonly used pressures with the ranges of (0-150 psi, 0-1500 psi, 0-3625 psi, 0-5800 psi, 0-8700 psi) [10 bar, 100 bar, 250 bar, 400 bar, 600 bar]
- User definable measurement units (psi/bar) for convenient and familiar data readings
- Ports: MNPT, SAE, couplings (push-button, sleeve operated, EMA3) to make plumbing and connecting easier and faster
- Corrosion resistant materials for challenging environments
- Sensor also provides temperature values
- User selectable scan and transmit rates (mode dependent). Currently 1, 2, 5, and 10 seconds. Refer to SCOUT Mobile for the up to date capabilities and modalities

Fig. 14.10- Wireless Pressure Sensor (Courtesy of Parker)

14.1.8- Excavator Experiencing Low Power

Figure 14.11 shows a trenching *excavator* that starts to experience slower performance.

Fig 14.11 – Trenching Excavator

❖ **Step 1: Review Safety Instructions:**
- Safety instructions for the machine/work environment were reviewed.

❖ **Step 2: Review Machine History:**
- By reviewing the machine history, no recent changes were reported.
- Checking the service manual shows that the actual duty is 25% longer in time as compared to the design duty cycle.

❖ **Step 3: Identify Main System Fault:**
- Actuator Stop to Move.

❖ **Step 4: Apply the System-Level Troubleshooting Chart:**
- "T-System-05: Low Power System" →
- Air filter of the engine was found workable in good shape and is not clogged.
- Power transmission elements were found ok, no slippage, worn or broken elements.
- "T-Pump-05: Low Pressure at the Pump Outlet" → no faults were detected.
- "T-Cylinder-01: Cylinder Troubleshooting" →
- "T-Seal-01-Seal Troubleshooting" →
- Piston seals were inspected and found to have axial cuts and particles embedded in the seal material due to abrasive contaminants.

❖ **Step 5: List Suspicious Components:**
- Skip.

❖ **Step 6: Perform Preliminary Check of the Suspicious Components:**
- Skip.

❖ **Step 7: Apply Detailed Check of the Suspicious Components:**
- Skip.

❖ **Step 8: Fault Evaluation Decision for Repair or Replacement:**
- Piston seal was replaced.
- System filter cartridge was replaced.
- System was drained, flushed, and filled with clean oil to specs found in the service manual.

❖ **Step 9: Startup and Testing:**
- Completed.

❖ **Step 10: Future Considerations and Documentation:**
- Perform routine maintenance on time and frequent fluid analysis is advised.

14.2- Case Studies Using Analytical Fault Detection Methodology for Industrial Applications

An *Analytical* Fault Detection Methodology requires highly experienced personnel. Time to find the fault depends on the experience of the expert in hydraulic control technology and the application of the machine.

This section was graciously provided to this textbook as a Courtesy of "CFC Industrial Training".

Robert Sheaf
President
CFC Industrial Training
A CFC-Solar company
7042 Fairfield Business Dr.
Fairfield, Ohio 45014
513-874-3225

14.2.1- Assembly Machine

Problem: Retract Coil Burn-Out (Fig. 14.12):
Workers at a private contractor that makes torpedoes for the U.S. Navy were having a problem with an *assembly* line directional valve coil that would burn out often, causing production issues. The valve, a D05, three-position, spring-centered, 115-volt AC explosion-proof design, continually lost the retract side "air gap" solenoid coil. The assembly line took six months to install and debug. Once it was turned over to production, this problem started showing up. The OEM sent a new valve, thinking there must be a problem with the original unit. It didn't help. The contractors' electrical engineer suggested they interchange the "A" and "B" ports to the cylinder and do the same with the two coils. The problem still persisted. Whichever coil controlled the retract function would eventually fail. The pump delivered 20 gpm, and the valve was rated for 26 gpm. What was causing the problem?

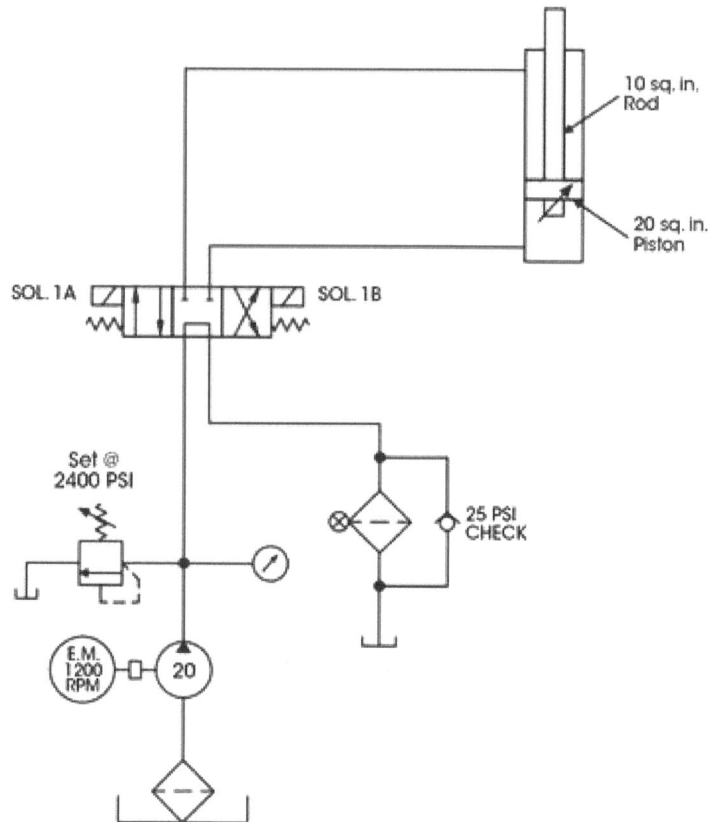

**Fig 14.12 – Circuit Diagram for Assembly Machine
(Courtesy of CFC Industrial Training)**

Solution:
The D05 explosion-proof valve that was used on the Navy torpedo line had an air gap desgined coil. This style of coil needs to fully shift for the high in-rush amps to drop to the low holding amps. This shouldn't have been a problem, especially for new valves. However, because the cylinder has a rod to cap end area ratio of 2:1, the returning oil from retracting the cylinder would force 40+ gpm flow through the directional valve that's only rated for 26 gpm. Many valve spools will try to move to the center position when their maximum flow rate is exceeded. This will cause the air gap solenoid to move off of the fully shifted position, increasing the holding amps causing the coil to fail. The OEM replaced the D05 valve with a larger valve to solve the problem.

14.2.2- Forging Machine

Problem: Overheating on a Large Forging Press (Fig. 14.13):
A *forging* and stamping company had problems with overheating on a press. The company had several identical presses, and only one had been overheating for some time. The employees added a larger heat exchanger a few months back, but they recognized that this was not the correct solution since the other identical presses worked well.

The circuit shown has been simplified; the unit had three pumps working together, several cylinders connected in parallel, and several accumulators all tied together. All the machines had identical kidney loop cooler and filtration systems that were proven to be working properly.
The only feedback we could get on the over-heating machine was the loud banging noise when the machine was shut down. This was caused by the accumulator circuit discharge rate not being throttled down properly. But this shouldn't have caused an overheating problem. The employees found that one pump's compensator was set slightly higher than the safety relief. When its pressure was set properly, the overheating problem lowered a few degrees, but that wasn't causing the main overheating problem.

The overheated oil was coming from the main manifold. They screwed the flow control rod on the 16-mm cartridge all the way in to lock the moving element in place, thinking it was popping open each cycle. It didn't solve the problem. The main safety relief was set properly. Any idea what could be causing the problem?

Solution:
When the workers couldn't find a problem causing overheating on the press, they started removing each component one by one hoping to find something that failed. When they got to the accumulator unloading valve, they found the sleeve and moving poppet were eroded away, allowing 335-bar fluid to discharge to the tank. This would heat the volume of oil leaking back to the tank up approximately 30 to 35 F. They pulled the sleeve out of the manifold and also found that the steel surface around the "B" side of the sleeve was badly eroded as well, but not to the point it would cause a problem.

High-velocity 335-ar oil discharging to a low-pressure tank line can cause cavitation. The fix was to add a 5-bar check valve to the tank line after the logic valve when it leaves the manifold. This back pressure is enough to prevent the cavitation bubbles from forming and eliminating the erosion problem.

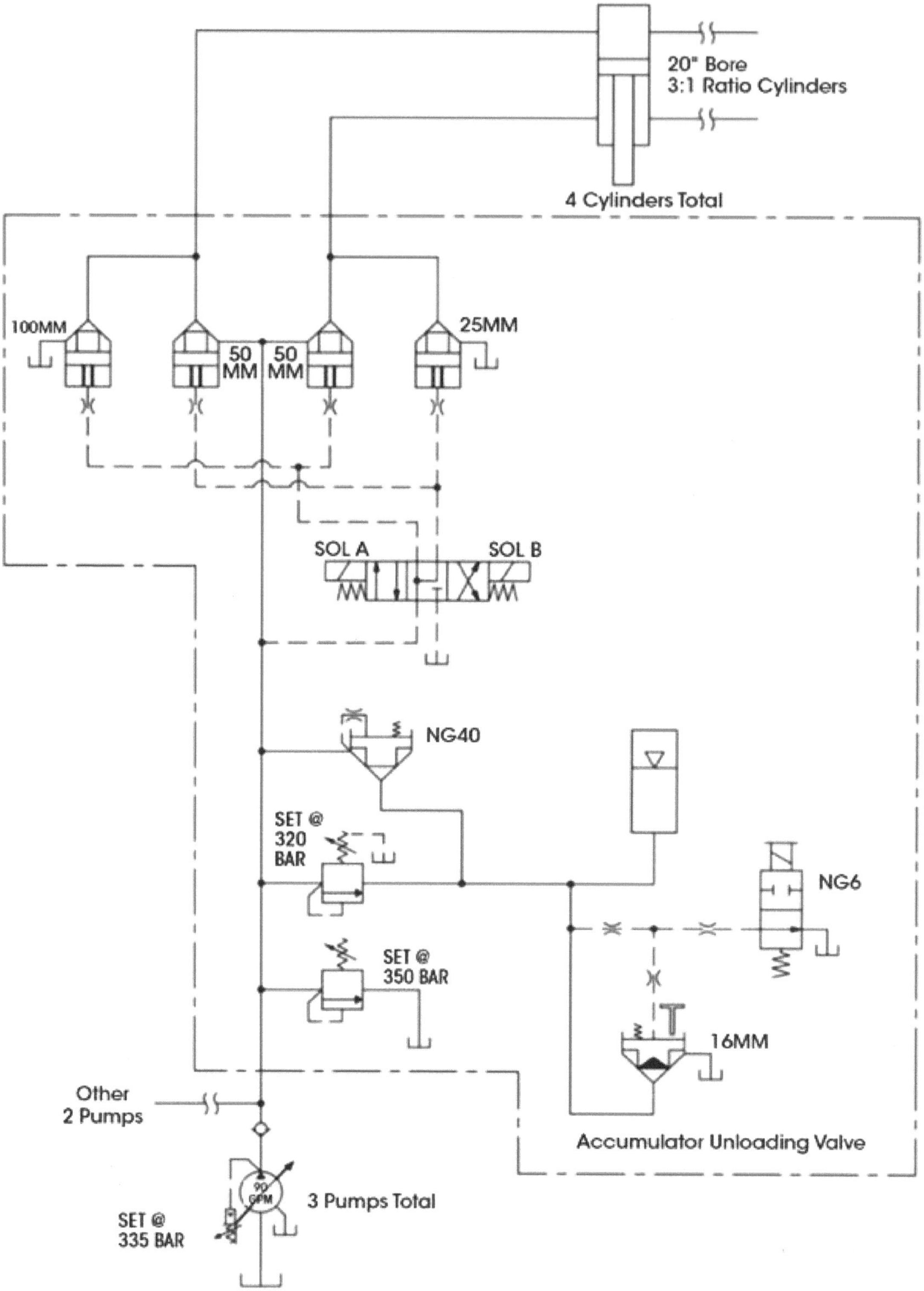

**Fig 14.13 – Hydraulic Circuit Diagram for Large Forging Press
(Courtesy of CFC Industrial Training)**

14.2.3- Hydraulic Press

Problem: Overheating of Press with Heated Platen (Fig. 14.14):
A small heated platen *press* for an aerospace parts manufacturer would close and hold pressure for 1 to 3 hours based on the part being molded. When the system was first installed, it developed trouble with the oil overheating. The employees made a design change, installing a 5-psi check in the pump case drain line back to tank and a ½ psi check going to the main return line. (See the circuit showing the final design.)

This modification worked just fine. When the press was in the clamped mode, the case drain would pass through the filter and heat exchanger, cooling the pump case drain during the holding phase. If the cylinder was cycled, it would cause flow and some back pressure in the main tank line and the back pressure would hold the ½ psi check closed, causing the pump case flow to divert directly to the tank over the 5 psi check. A 5-psi check was used since the shaft seal of the pump was limited to 10 psi.

Every so often, the press would start overheating during the clamped cycle. The employees would lower the pressure setting on the safety relief to flush any contamination that may cause it to leak and reset it back to 3000 psi. They would change the filter element, which looked dirty, and check the water connections on the heat exchanger for flow or corrosion. What could be causing the problem, and what would be the solution?

Solution:
When pressure-compensated pumps idle for long periods of time, the case flow can cause the reservoir oil to overheat. Using a 5 and ½ psi check valve circuit allowed the case flow to be cooled when the main system is static, and the pump is compensated. When the designer installed the modification to cool the case drain, he should have connected the outlet of the ½ psi check should have been to only flow through the cooler and not the 3-micron filter. The 15 gpm filter was undersized to begin with since the return flow from the cylinder retracting was 24 ½ gpm. As the dirt and contaminants accumulated in the filter, the pressure drop increased. As soon as the pressure drop exceeded 4 ½ psi, the case drain flow was returning to the reservoir through the 5-psi check heating up the oil.

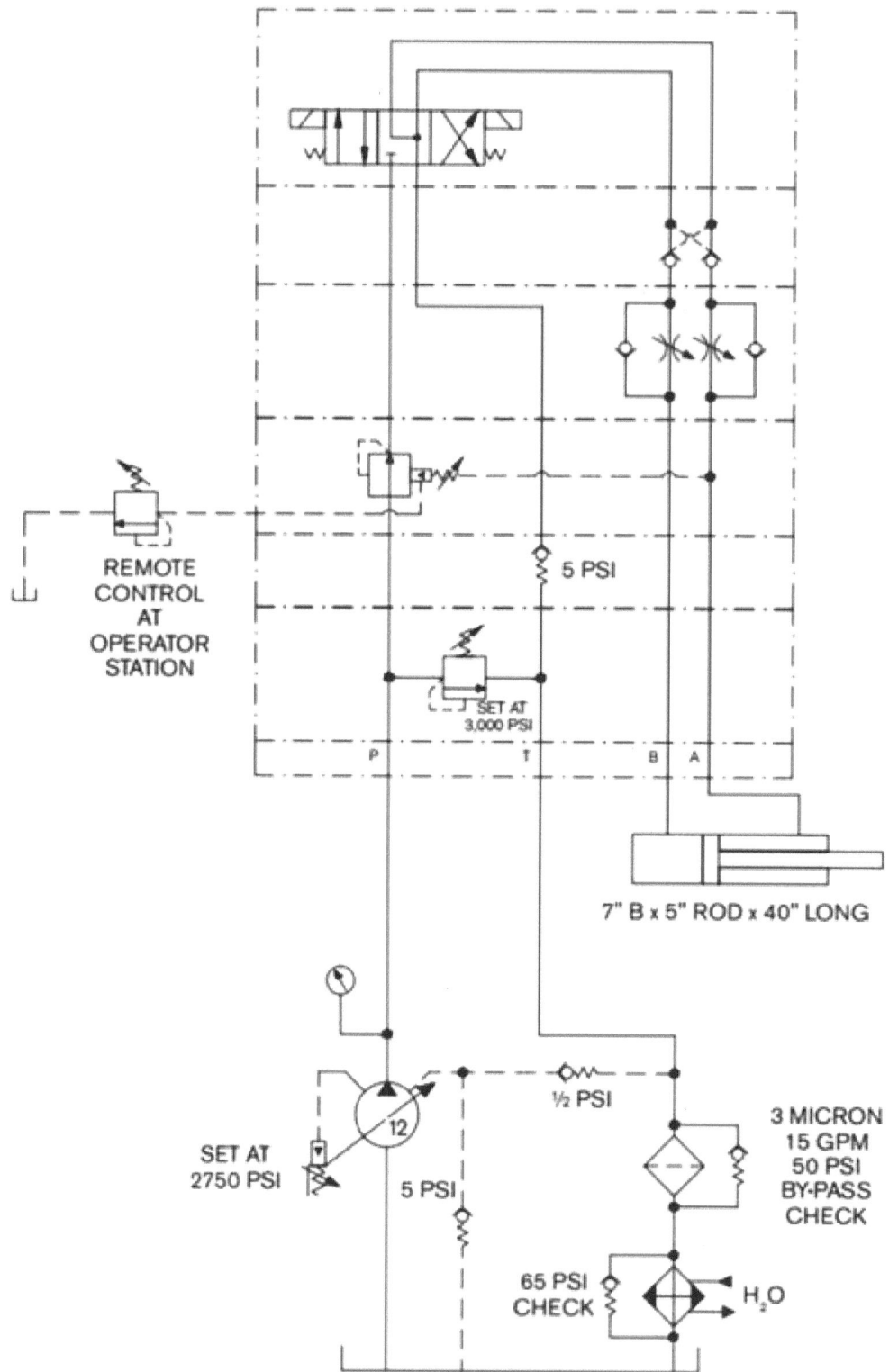

Fig 14.14 – Hydraulic Circuit Diagram for Heated Platen Press
(Courtesy of CFC Industrial Training)

14.2.4- Scrap Steel Winder

Problem: Scrap Steel Winder Problem Slowing Down (Fig. 14.15):

Workers at a coil *steel-processing* mill received rolls of stainless steel from its parent company to split in half and trim the edges approximately 1″ to 2″. They recoil the sheets and ship them to their OEM customer that fabricates stainless steel refrigerators. They have a hydraulic motor-driven unit that coils the scrap edges of the coils onto drums. This operation needs to keep up with the speed of the splitter, or the scrap starts to "bunch up," which requires the workers to stop the line and remove the excess manually. The hydraulic unit was over 20 years old and needed updating. A local hydraulic distributor built them a new unit, and the mill did the installation. During commissioning, they found that the new unit could not keep up with the slitter line speed as the scrap coil increased in diameter. The unit was upgraded from a fixed-volume open-circuit design to a closed-circuit pressure-compensated pump design to improve energy efficiency.

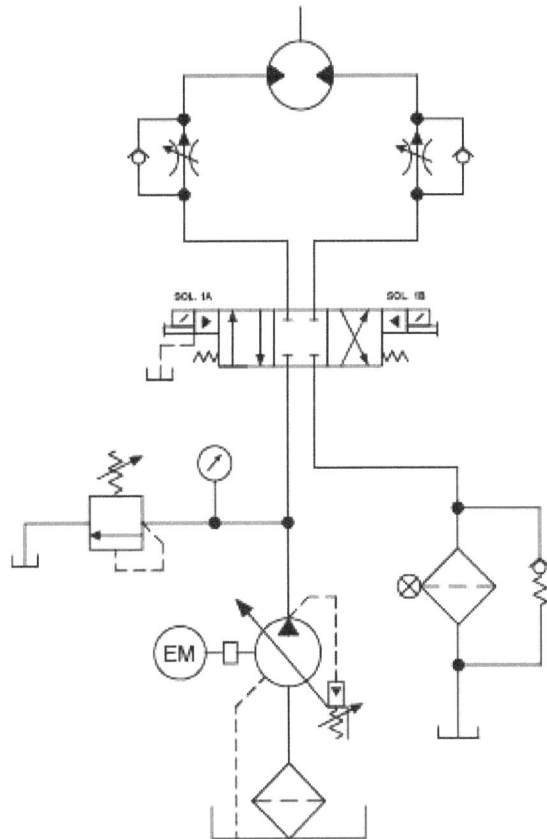

Fig 14.15 – Hydraulic Circuit Diagram for Scrap Steel Winder (Courtesy of CFC Industrial Training)

The only other difference was the use of a flow-control module under the pilot-operated directional valve. The old unit had line-mounted flow controls. The builder convinced the mill that it must be the mill's existing motor leaking more flow to the case, as the pressure increased when the scrap steel diameter increased, causing the motor to slow down. They then replaced the hydraulic motor with a new one, but they still had the same problem. Any idea what the problem could be?

Solution:

The reason the hydraulic motor on the slitter slowed down was the use of non-pressure-compensated flow controls in the module mounted under the directional valve. They missed the fact that the original flow controls were pressure compensated. With non-pressure-compensated flow controls, the flow they pass decreases as the pressure drop across them decreases. The pump compensator setting stayed consistent, while the motor pressure increased as the scrap coil diameter increased, reducing the flow. Replacing them with pressure-compensated flow controls solved the problem.

14.2.5- Gold Mine Crusher

Problem: Gold Mine Crusher with Erratic Pressure (Fig. 14.16):

A hydraulic power unit operating a *rock-crushing* machine had erratic pressure problems, causing the crusher to stall when the pressure seemed to relieve at a lower pressure than needed. The machine was over 30 years old and was somewhat quite simple in design. The circuit attached shows the main relief valve remotely controlled so the operator could reverse the motor's rotation with the directional valve and adjust the pressure higher than normal to clear jambs. Once the jamb was cleared, he would reset the pressure back to the normal setting and shift the directional valve back to rotate the crusher motor in the normal operational direction. They determined that the remote-control relief valve was worn badly and needed to be replaced.

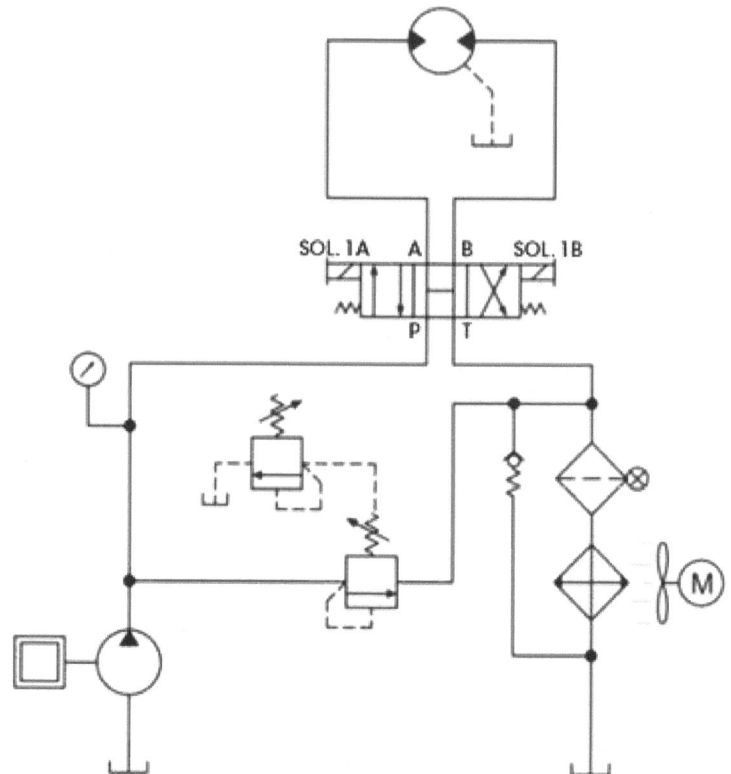

**Fig 14.16 – Hydraulic Circuit Diagram for Gold Mine Crusher
(Courtesy of CFC Industrial Training)**

Their storeroom had the same brand of relief with the same spring range, but with just one letter difference in the model number. They installed it in the manifold, adjusted the valve from fully open to the number of turns closed used by the old, similar valve with the same spring rating, and then started the system up. It seemed to run fine until the crusher jammed and blew the side of the pump out. Luckily, no one got hurt. What could have caused the problem?

Solution:

The gold mine power unit where the side of the pump blew out had a relief valve installed that had the inlet and outlet reversed, taking the pilot relief out of the circuit. Without a drain, the main relief would not open, causing an unwanted pressure spike that blew the side out of the pump. Sun, Parker, and Eaton all over the years have produced relief valves where the pressure inlet comes into the sides and out the bottoms, as well as units that have the inlet in the bottom and out the side. It's so important to verify which ports on the manifold or housing are inlet and outlet, and that the screw-in cartridge matches the same configuration.

14.2.6- Training Stand

Problem: Cylinder Stuttering on a Technical School Trainer (Fig. 14.17):

A community technical school received funding for a hydraulics program. They had nice hands-on-training stands in storage and did a nice job setting up their hydraulics lab. They talked a Physics instructor into teaching the class, and things were going fine until he had the students plumb up the circuit shown. The cylinder would stutter badly when it was extending and run smoothly when it contacted a load spring. The professor thought it was just a case of air in the cylinder, but the students could not see any air spitting out when they bled both ends of the cylinder. They were thinking the cylinder was damaged inside and should be replaced. The professor moved the students over to a spare hands-on trainer and found it also stuttered using the same circuit.

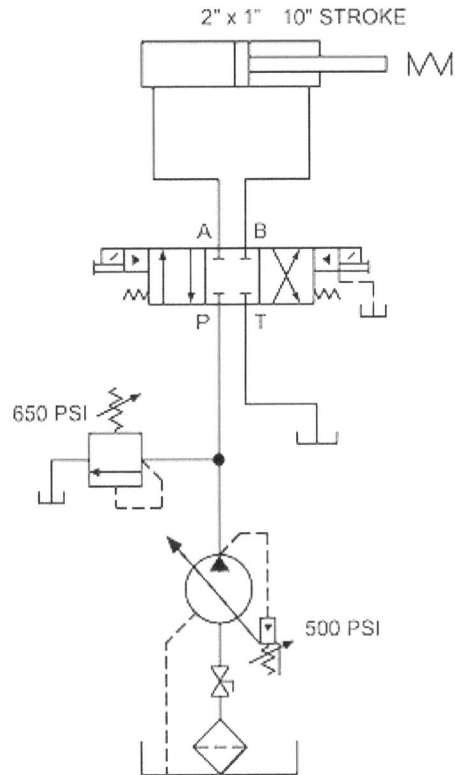

Fig 14.17 – Hydraulic Circuit Diagram for Training Stand (Courtesy of CFC Industrial Training)

The circuit used a pressure-compensated pump set to 3 gpm; a pilot-operated, closed-center directional valve with internal pilot; and external drains and a cylinder. The cylinder would extend 6″ before it contacted a die spring, which would simulate a load, causing the pressure to build up as the spring compressed. The flow meter indicator bounced up and down when the cylinder stuttered but held steady when the load spring was engaged. The glycerin-filled pressure gage dropped to a low reading until the load spring was compressed.

Solution:
The stuttering cylinder without a load only needed very low pressure to extend. Even though the pilot-operated directional valve had a closed center, it still needed at least 75 psi to keep it shifted. The cylinder would start to extend at a pressure lower than 75 psi, causing the directional valve to un-shift, momentarily stopping the cylinder. As the valve un-shifted, it blocked the flow, causing the pressure to build higher than 75 psi, shifting the valve open again. This pressure fluctuation was too quick for the gage to register, but it showed up in the cylinder and flow meter. Putting a little backpressure in the tank line or increasing the pump flow solved the problem.

14.2.7- Log Splitter

Problem: Aerated Oil on a Splitter (Fig. 14.18):
I loaned my *log splitter* to a friend whose tree fell after a storm. The tree was a blue spruce evergreen about 30-ft. tall. It surprised me that the tree's root system, after 20 years of growth, barely penetrated the ground but seemed to be spread out over a 10 to 12-ft. radius being only about 2 to 6-in. deep.

He had cut the tree up in short lengths that would fit into his free-standing fireplace and needed to split and stack the logs. He borrowed my splitter and complained that it slowed down when splitting the logs—slower than it had when he helped me in the past.

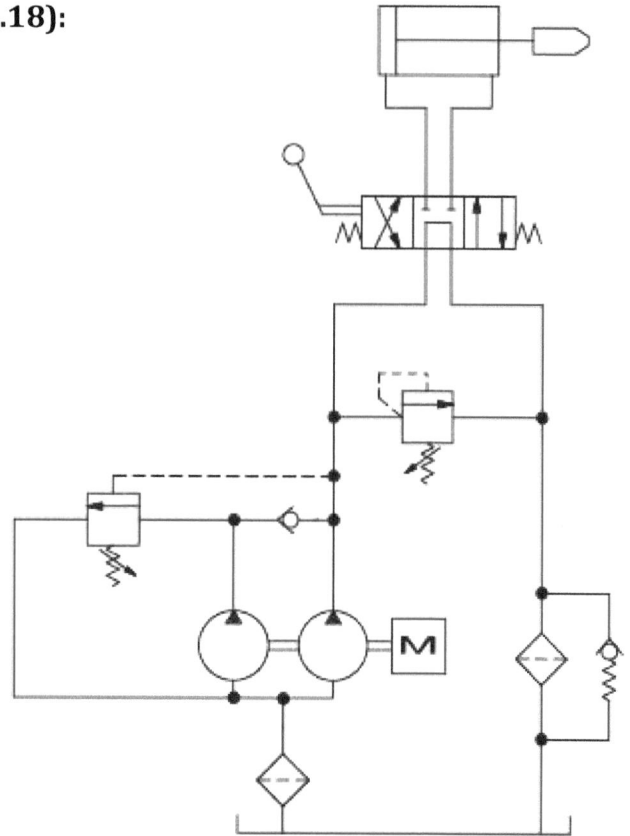

**Fig 14.18 – Hydraulic Circuit Diagram for Aerated Oil
(Courtesy of CFC Industrial Training)**

I explained to him that the splitter had a "Hi-Low" pump system with a new cylinder and that it would slow down when it encountered logs that needed pressures higher than 600 psi (see the circuit). He felt this wasn't the problem, so I went over to his house and looked at the splitter. He was correct; the splitter encountered the log, slowed down, but eventually split the log. I was sure it was the pump failing, but without a gage or flow meter I removed the pump and took it to my shop for testing. To my surprise, the pump tested out just fine.
Any idea what the problem could be?

Solution:
The log splitter problem of taking too much time to pressurize and slowing down the cycle time was caused by the oil having excess air mixed with it. The return line was returning oil above the fluid level, aerating the oil. The air slowed down the ability of the oil to pressurize quickly. Filling the reservoir with more oil to proper level solved the problem.

14.2.8- Walking Beam

Problem: Walking Beam Cylinder Drifting Up (Fig. 14.19):
A processing plant in Pakistan had a *walking-beam* system that would move coils of finished product to shipping for processing. Several "pot-belly" cylinders would lift long, horizontal rails with "V" notches cut in the top surface.

A large shaft was inserted through the middle hole of the coils and extended out past the coil sides, and would nest in the notched rail, lifting the coils up and above fixed "Y"-shaped supports. Once lifted, two other cylinders would move the rails horizontally, approximately 8 feet. Then the "pot-belly" cylinders would retract, lowering the coil shafts into the fixed supports. The lifting rail was retracted 8 feet, lifted and horizontally advanced the coil 8 more feet towards shipping. One large, pilot-operated directional control valve would move 8 cylinders.

After an annual shut down, where the long hoses and directional valve were replaced, the cylinders started drifting up, contacting and lifting the horizontal rails, which caused a problem. The employees checked the tank line check valve, and it was working fine. They suspected the new directional valve since they had to switch the solenoid wiring to get the correct action.

Any idea what the problem is?

Solution:
The lifting cylinders were drifting up, so this ruled out bypassing pistons. It turned out the employees had switched the main cylinder feed hoses. This explained why the solenoid wires had to be switched.

In addition, with a spool-type valve that has a blocked "P" port, leakage from the "P" port to both "A" and "B" ports exist, causing a pressure increase in any blocked cylinder ports. The pressure builds up close to half of the system pressure, which was enough to cause the cylinders to drift.

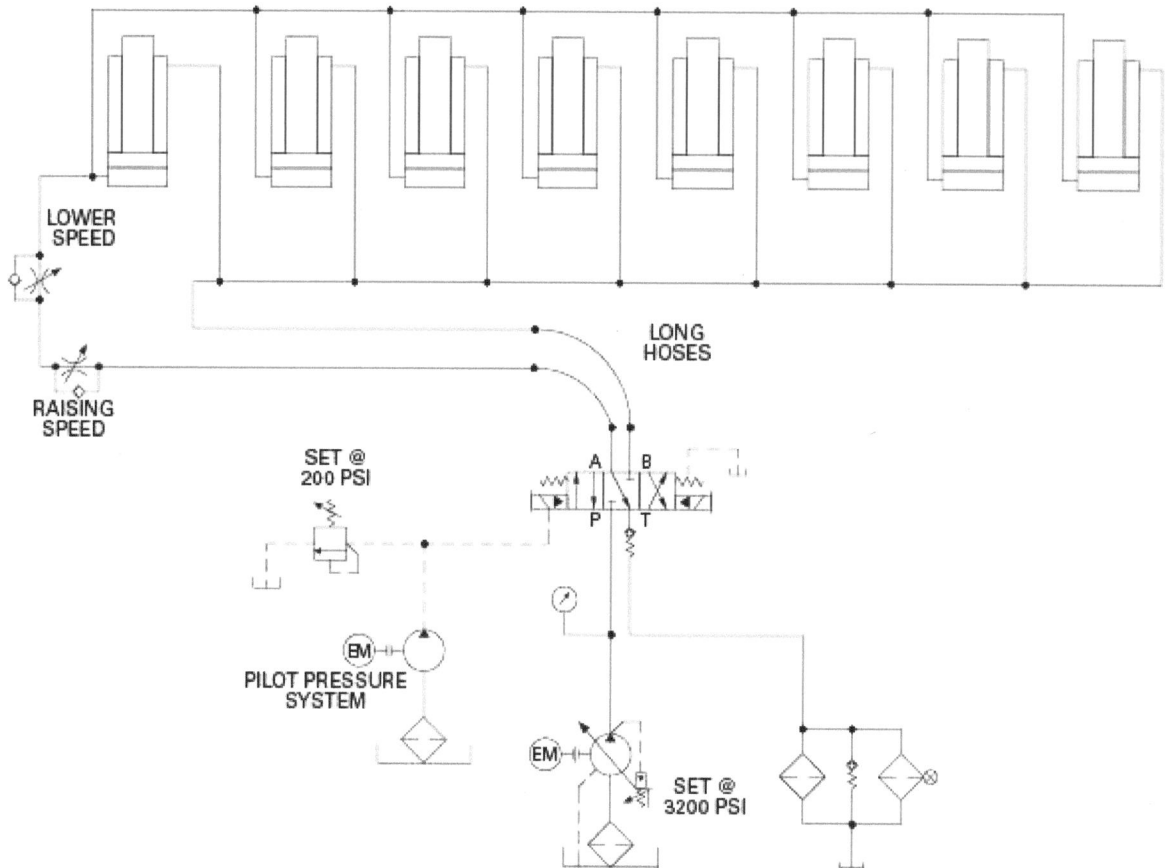

**Fig 14.19 – Hydraulic Circuit Diagram for Walking Beam Cylinder
(Courtesy of CFC Industrial Training)**

14.2.9- Load Sense Pump

Problem: Improper Operation of Load Sense Pump (Fig. 14.20):
A hydraulic distributor hired a sales engineer for a new territory they were moving into. The first customer the engineer visited requested a quote on a hydraulic system. The customer provided the following requirements to press two large forgings together:

- 5" bore cylinder, 3.5" rod & 36" long
- DCV, 3-position with a return "kick back to center" feature, manually operated
- Flow controls on both cylinder ports
- Return line filtration
- A pump that delivered 30 gpm at 2450 psi (not piston-type)
- No heat exchanger, if possible (The customer was going to build a larger reservoir than normal.)
- Electric motor had to be 230 volt, 3-phase, 60-cycle, C-face with coupling.
- All parts needed to ship loose for the customer to mount and plumb on his machine.

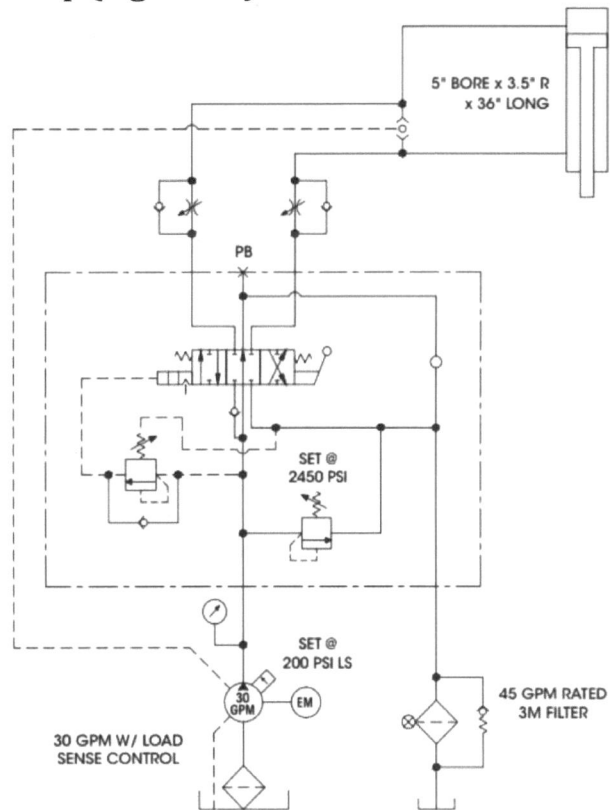

Fig 14.20 – Hydraulic Circuit Diagram for Load Sense Pump (Courtesy of CFC Industrial Training)

The sales engineer designed the circuit and quoted the customer a price to deliver all the components. The customer installed the parts using the circuit supplied. The sales engineer then got a call from the customer stating that the pump was overheating. It also seemed as if the *load sense* worked when retracting, but when extending at no load, was always at 2450-psi system pressure. The return line screw-on filter element was coming loose, and the customer had to keep tightening it. Any idea what's causing the problems?

Solution:
When using load-sense circuits, it's important to locate the sensing line where it will always sense the load pressure required to move the cylinder. The shuttle valve location on the circuit should have been between the DCV and the flow controls. When the cylinder is extending, the rod pressure can be twice the cap pressure due to the meter-out flow control causing the load sense to go above the main relief setting. Also, the filter was undersized, since a 2 to 1 ratio cylinder will return twice the pump flow at full speed when retracting, causing the element's thread stud to stretch and eventually blow off.

14.2.10- Food Processing Equipment

Problem: Drifting and Lurching Cylinder on Food Processing Equipment (Fig. 14.21):
In the *food* and *pharmaceutical* processing industries, leaking actuators can contaminate products and increase waste, not to mention the possibility of contaminated product getting out to the public. Generally, food-grade hydraulic oil is utilized, but this could affect the taste, and some consumers could be allergic to the oil.

Hydraulic cylinders in these industries are installed with an additional outboard rod-cap seal as an additional safety measure in case the main rod seal starts to leak. This is commonly referred to as "drain-back." In effect, it makes two rod seals, in series, with a clear plastic tube draining the area between the two caps. The clear plastic tubing would drain any leaking oil back to tank before it had a chance to drip on the product. Drain lines like these are one of the rare examples where terminating drain lines above the fluid level should be used. These drain tubes are routinely checked for signs of oil leakage so the maintenance technician can schedule a cylinder replacement at a shutdown.

The cylinder in question was located approximately 40 ft. above the hydraulic power unit (HPU). The unusual drain line with 5-psi checks would ensure fresh, clean oil migrating to the cylinders and return oil making it back to the tank. The D05 (NG10) proportional directional valve was mounted on the HPU.

The food industry is very sensitive about intellectual property and system designs. When I was asked to troubleshoot a problem that developed with one of the cylinders used on a new meatball manufacturing process, they had large covers around the process so that the only thing I could see when I arrived was the HPU unit. I was not allowed to see the complete circuit design, just the attached part of the circuit (see illustration).

The problem was two-fold. If the line stopped for more than 15 to 20 seconds, the cylinder would drift in the retract direction and trip a fault sensor. Moreover, every time they shifted the proportional valve, it would cause the cylinder to "jump" as if there were air in the system. They felt they cycled the system sufficiently to remove the air, but it continued to jump and lurch when actuated. They unplugged the electrical signal to the proportional valve in an attempt to stop the drifting, but the cylinder would still drift.
Any thoughts on what the problem might be?

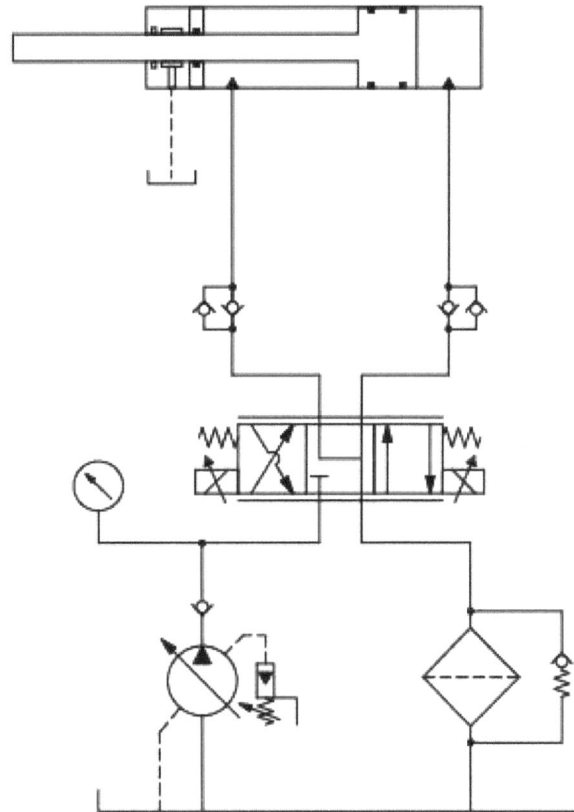

**Fig 14.21 – Hydraulic Circuit Diagram for Food Processing Equipment
(Courtesy of CFC Industrial Training)**

Solution:
Drifting: The problem causing the cylinder to drift and retract was the weight of the 40-ft columns of oil, which opened the return checks and created a vacuum on both sides of the cylinder.

The return 5-psi checks needed stronger springs to keep them from opening. Oil weight is approximately 0.4 psi per foot times 40 ft, or 16 psi, or approximately 30"Hg vacuum. The larger cap-end area of the cylinder with a vacuum caused it to drift.
Lurching: Proportional coil armatures will jump if air is in the coils. Many proportional valves have coil air bleed screws. A lot of designs would install 5-psi checks in the tank line before the filter to help facilitate the removal of the trapped air.

14.2.11- Hydraulic Elevator

Problem: Filter Issue in a Hydraulic Elevator (Fig. 14.22):
A well-known actor had a small *elevator* installed in his mansion that sits on top of a Tennessee mountain. He wanted it to lift two people up 20 feet to a roof-top balcony. Since there was not a roof to mount the typical electric drum and drive, a hydraulic lift was installed. The circuit was simple, using a vane pump running at 1200 RPM to keep the noise level to a minimum.

The hydraulic system was designed so the maximum pump flow determined the lifting speed and a flow control determined the lowering speed. The control system was designed to use 24 VDC with a battery backup in case of a power failure. Since the valve was directly operated, battery power could be used to shift the valve and lower the elevator down from the roof top. It is questionable if the designed circuit met all state and local safety standards.

After a year or so, each time the unit was started, it rumbled for about two minutes and then settled down to a normal running sound. The system designer was nowhere to be found, so a local repair shop was contacted. The employee concluded the unit was losing its prime. Since there were no signs of external leakage where the air could be getting in, the repair shop employee convinced the owner that a check valve installed in the inlet line would solve the problem.

The unit worked well until one day the electric motor overloaded, tripping while lifting the elevator, causing a jerk. Then a loud bang was heard, followed by the elevator lowering to the bottom, even though the raise button was selected. The passengers were only four or five feet up on their way to the roof when this happened, and luckily no one was hurt.
It was found that the externally accessible inlet filter's top cover plate had failed and blew oil everywhere.

What caused this to happen, and how could it be resolved?

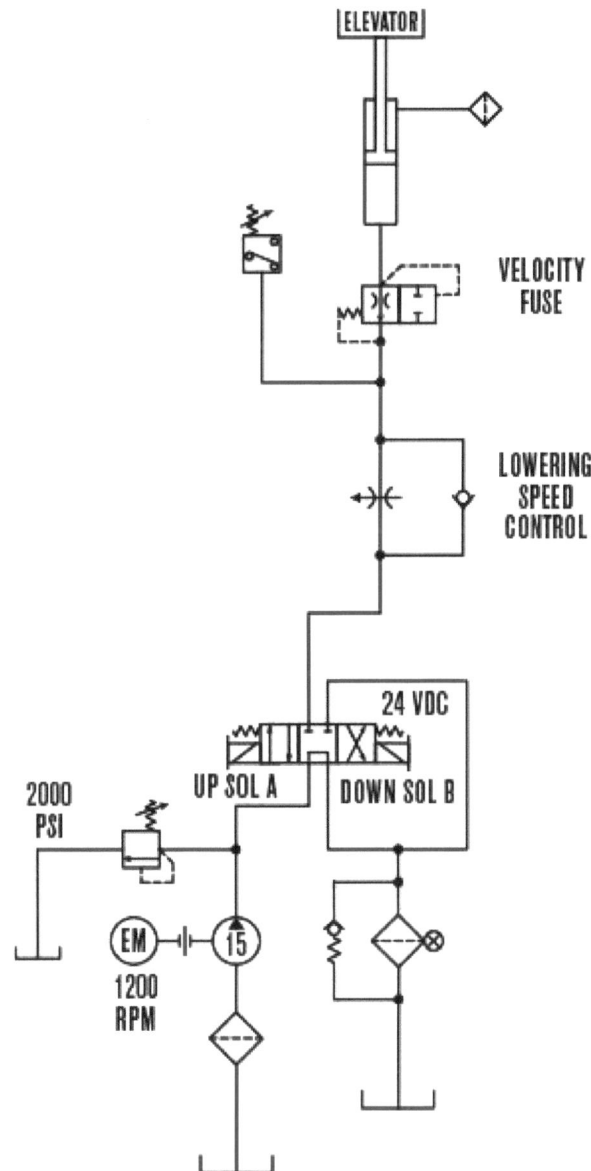

**Fig 14.22 – Hydraulic Circuit Diagram for Hydraulic Elevator
(Courtesy of CFC Industrial Training)**

Solution:

The mansion elevator that blew its intake filter top off was caused by a failure of the electric motor with the directional valve still shifted. The cap end of the lifting cylinder was connected directly to the stopped vane pump, allowing the backward flow of oil, supporting the lift to build up pressure in the inlet line of the pump.

The check valve that was installed to solve the priming problem would not allow the oil flow to drain back to the tank. The weak point in the inlet line was the intake filter's lid failing, spraying oil everywhere.

14.2.12- Polishing Machine

Problem: Maximum RPM Couldn't be Reached (Fig. 14.23):

A casket manufacturer has a machine that polishes the stainless steel exterior on finished caskets to a fine "brushed" look for one design that's very popular. The *polishing* head is powered by a "servo valve" hydraulic system that allows them not only to vary the speed of the turning brushes but also maintain the force exerted on the brushes. The servo valve electronic controls incorporated a feedback signal from the hydraulic motor to assure a constant RPM speed for the brushes regardless of the pressure applied.

This process has been in place for 10 to 15 years, and other than obvious maintenance fixes, it worked well. The workers did a good job of maintaining the ISO cleanliness code and temperature of the hydraulic oil. Oil samples always came back reflecting the good maintenance practices they had in place.

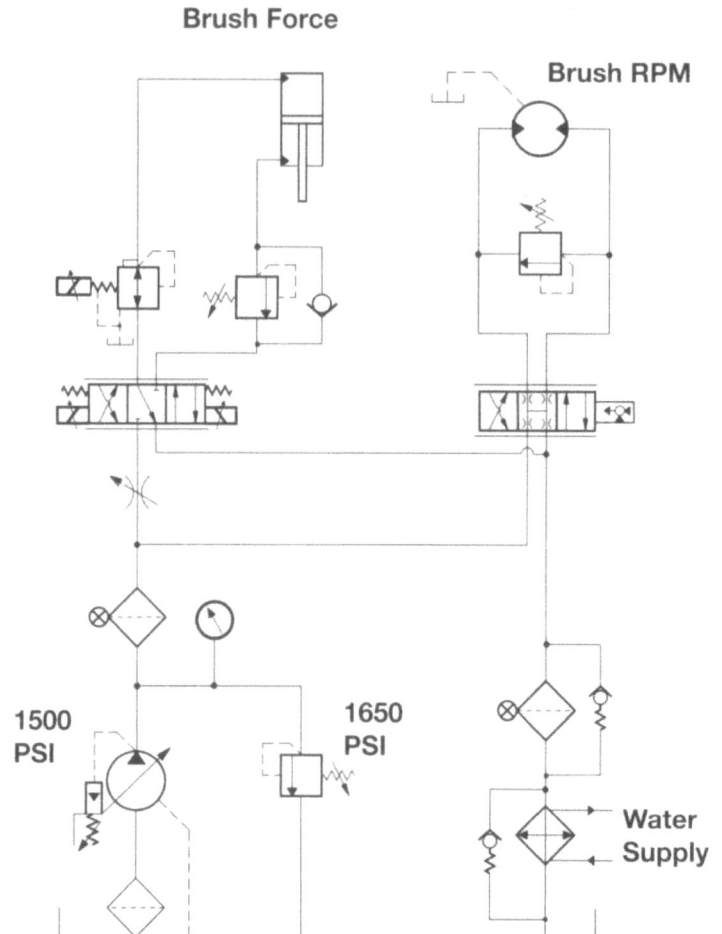

**Fig 14.23 – Hydraulic Circuit Diagram for Polishing Machine
(Courtesy of CFC Industrial Training)**

However, a problem developed where the maximum brush RPM's could not be reached and maintained, even though the circuit for advancing and applying pressure worked just fine. (See the accompanying circuit.)

The coils of the servo valve were wired in parallel, and when the resistance of each was measured, they both appeared to be fine. The workers checked the signal coming out of the amplifier card controlling the servo valve, and it also seemed fine. They then replaced the servo valve with a new one after checking the continuity of the cable going to the valve.

The system would work fine part of the time but not consistently. Any idea what could be wrong?

Solution:
Servo valves have two coils that control the tilt of the torque motor. When wired in parallel and properly wired in relation to their polarity, the coils can produce enough force to fully stroke the valve. However, if one coil is "open," the valve only has half the force stroking the torque motor and cannot fully open, reducing the maximum flow capacity and slowing down the response time of the valve.

The fact that it worked fine part of the time would indicate the hydraulics was not the problem. I found that if the cable feeding the valve was moved to a certain position and held there, the system worked fine. Closer inspection of the cable, where it was soldered to the connector terminal, found the solder joint of one wire was a "cold joint" and the wire separated from the terminal when the cable was bent one direction. Re-soldering the wire to the terminal solved the problem.

14.2.13- Power Unit

Problem: Pressure Losses:
When I was a manager of a large hydraulic *power unit* (HPU) fabrication shop, we had a strange problem while testing a large HPU. It consisted of a large vane pump, a pilot-operated relief, and a 5-bank 008 manifold with 3-position blocked center directional valves with pressure reducing modules. All the A and B ports contained pressure gauges. The HPU had a separate off-line filtration and cooling circuit, which we commonly refer to as a kidney loop. The pump was driven by a 100-hp, 1800-rpm electric motor - a large HPU, but a fairly simple system. The system pressure was set to 2250 psig with an output flow of 75 gpm. The pump would run for 30 to 60 seconds at a time under pressure, but then all pressure would quickly drop to near 0 psig for 2 to 3 seconds. Pressure would then quickly jump back up to 2250 psig. This happened 15 to 20 times before the test mechanic called me for help. The unit piping had about Y2 NPT and Y2 0-ring connections. No Teflon tape was used, just white, paste-type pipe dope for sealing the NPT connections. We felt the only valve that could cause the malfunction was the pilot-operated relief, so we dis- assembled and inspected it. We did this at least four or five times and could not see any contamination or sticking problem. We were pressed to move onto other units on the test stand, so we drained the oil from the reservoir and removed its access panels, with the eventual plan of retesting it later. However, we noticed that the reservoir bottom was covered with typical construction debris and a rather large number of small beads of excess pipe dope. Any idea what was causing the problem?

Solution:
Many people do not allow the use of PTFE (Teflon) tape for sealing NPT fittings. However, excessive pipe dope during construction can lead to failure of pilot-operated relief valves when the pipe dope oozes through the control orifice. This causes the valve to open until the pipe dope exits, then the valve resets itself.
The unit was disassembled, cleaned of all excess pipe dope reassembled, tested with good results, and then shipped out to the customer.

14.2.14- Hydraulic Press

Problem: Cylinder Shuddering in Hydraulic Press (Fig. 14.24):

I was called out to troubleshoot an 800-ton *press* that molded large aerospace seals. The press would close at normal speed for approximately 2 ft., then slow to a creeping speed for another 6 to 8 in. to close the two-part mold. Subsequently, pressure built until it reached the required tonnage.
It held the tonnage for about five minutes, allowing the molded piece to cure. Afterward, the press retracted at a super-slow speed so that the piece could separate from the steel die. Then, after retracting 10 to 15 in., a valve shifted it into normal upward speed until fully open.

The press started stuttering on the normal downward stroke. It would start and almost stop several times. We determined it must be the main proportional flow control. The *super-slow speed*, as they called it, worked fine.

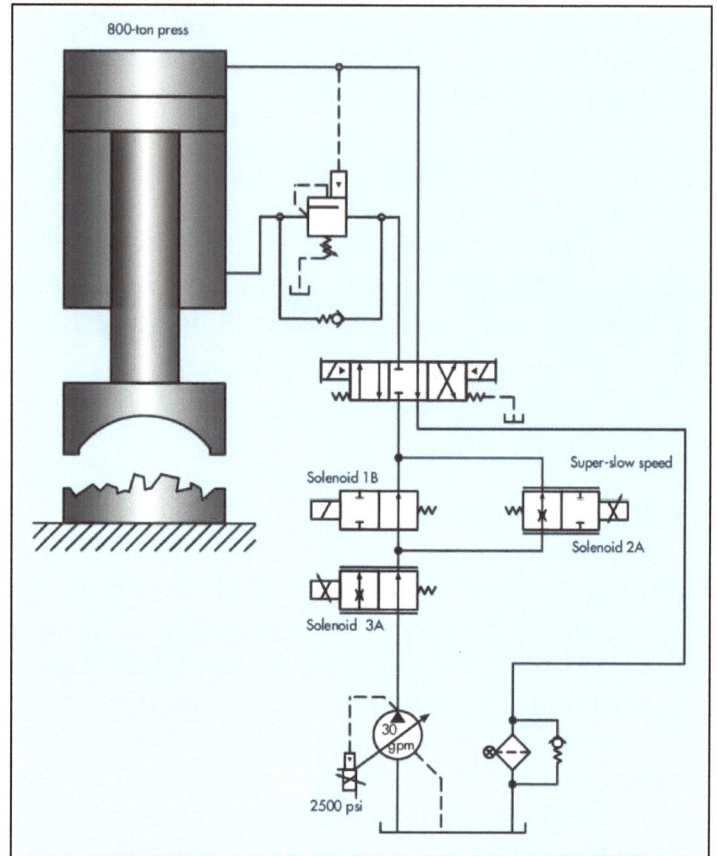

Fig 14.24 – Hydraulic Circuit Diagram for Cylinder Shuddering (Courtesy of CFC Industrial Training)

The diagram shows a simplified version of the hydraulic system mounted on top of the press, which is difficult to access. The super-slow valve is an Eaton IS0-3 mounted in the open, and valve no. 3 is a screw-in cartridge with a slip-on DIN coil mounted behind the pump assembly. The voltage and current to proportional valve no. 3 were measured at 12 V de and 4.6 A from the control panel at floor level. Disconnecting the signal to the valve allowed the valve to fully close. The typical command range of proportional valve no. 3 was normally 0 to 8 V DC at about 1 .3 A. At 0 V, the valve was able to fully close, and 8 V DC shifted it fully open.

A PLC code measured the speed of the ram and adjusted the proportional signal to maintain the programmed speed. A brand-new cartridge valve had been installed one week earlier, and a fluid sample indicated an ISO cleanliness code better than recommended. The press worked fine for a week. So, you would think if it was contamination, it would neither stutter nor fully close when the signal was lost. What do you think caused the severe stuttering?

Solution:

The shuddering on the 800-ton press was caused by the position of the proportional valve's solenoid coil was only halfway on the core tube. With the coil only partially over the armature, forces generated by high oil flow would overpower the magnetic force of the coil, causing the spool positioning error. Once the flow decreased, the flow forces became lower. This allowed the coil to again open the spool wider, and the process kept repeating.

14.2.15- Regenerative Circuit

Problem: Improper Operation of a Regenerative Circuit (Fig. 14.25):

An older *molding* machine was being used to debug new dies. The opening and closing time for the machine had increased to twice its normal high-speed closing rate. However, the tonnage when clamped was still working fine. The circuit used a Sun Hydraulics counterbalance valve as the regenerative unloader. Unfortunately, as is too often the case, the first thing the mechanic did was replace the pump, which didn't solve the problem. Next, the Sun counterbalance valve was replaced. The maintenance people felt if they adjusted it as high as it would go clockwise, the high-speed regeneration circuit should work all of the time, and they would see half the normal tonnage. Again, this did not fix the problem. They disassembled the check valves and relief but found nothing wrong. Any idea what could be causing the problem?

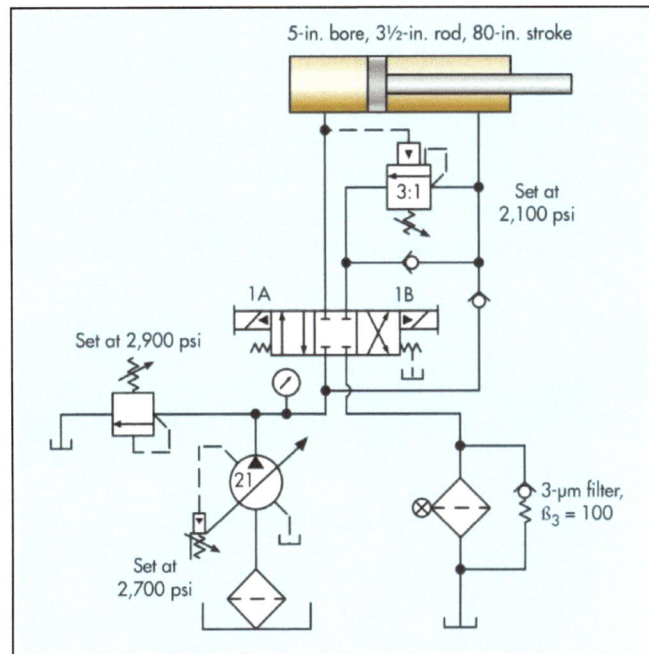

Fig 14.25 – Hydraulic Circuit Diagram for Regenerative Circuit (Courtesy of CFC Industrial Training)

Solution:

Regenerative circuit rely on the oil from the rod end of the cylinder to flow to the cap end, which causes the piston rod to extend faster than in a conventional circuit. If left in this condition., the annulus area on piston's rod side works against the area on the opposite side. The result is lower applied force from the cylinder. To get maximum tonnage, oil from the rod side needs to be connected to a tank line. The Sun counterbalance valve adjustment is opposite that of most pressure controls, which increases pressure when rotated clockwise and decrease pressure when rotated counter-clockwise. The mechanic needs to adjust the Sun counterbalance valve counter-clockwise to increase the pressure. Unintentionally adjusting the Sun valve to open at a very low pressure prevented the regeneration feature from activating.

14.2.16- Forging Machine

Problem: Damaged Valve on a High-Speed Forging Machine (Fig. 14.26):
A high-speed *forging* press used a large vane pump to cycle four different circuits. Three of the circuits were used for setup and eject functions and the fourth, shown here, provided a secondary stamping function. A vane pump was used because the constant pounding inherent to the application can cause problems with pressure-compensated pumps. The higher cost of a pressure-compensated piston pump can be another factor.

The main directional control valve became damaged and was replaced when a new die was being installed. A spare valve with the same model number except one letter was found in the shop's storeroom. The model number indicated the valve was configured for external pilot pressure, but the circuit called for one with an internal pilot.

Technicians removed a solid internal plug from the valve, which they figured would convert the valve to internally piloted. They installed this new valve and got the press up and running again. However, the operator said the machine didn't sound like it did before. He felt something was making a slamming or hard thumping noise each time the new valve shifted.

What do you think the problem was, and how would you correct it?

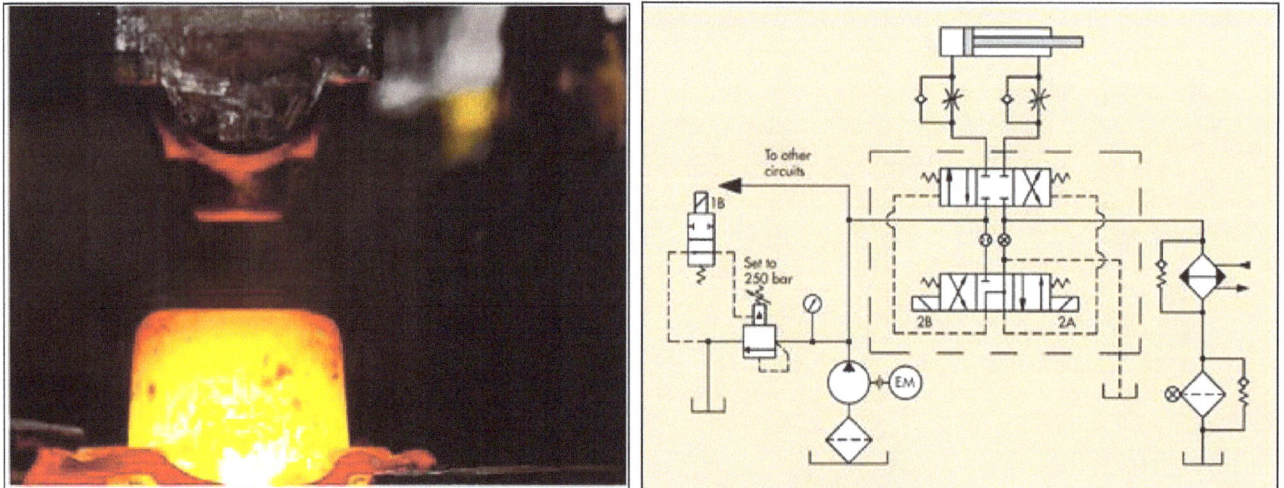

Fig 14.26 – Hydraulic Circuit Diagram for Forging Machine
(Courtesy of CFC Industrial Training)

Solution:
A pilot-operated directional valve can be externally or internally piloted. When these types of valves are ordered with an internal pilot connection, most manufacturers use a plug with an orifice to control the spool speed. It slows the valve down a few milliseconds and prevents the spool from slamming into the forged steel endcap, which would make a thumping sound. Installing the orifice plug from the old valve solved the noise problem.

14.2.17- Hot Dog Machine

Problem: Damaged Valve on a High-Speed Forging Machine (Fig. 14.27):

Packaging house has a *hot-dog machine.* It starts off with raw meat from large tubs dumped into grinders that feed the hot dog production line. After spices and fillers are added, the mixture is inserted into a casing material and cut to length. Then the raw hot dogs pass through an oven, and are subsequently wrapped and labeled. Finally, the packages are quick-frozen and packed into cartons for shipment. The large tubs of raw meat are lifted and dumped using a simple hydraulic system *(see schematic)* consisting of a cylinder that lifts the tub up to dump the meat into the grinders. Over time, operators noticed that if meat was fully loaded to the top of the tub, the cylinder could not lift it. The production crew changed the full-fill point to reduce the amount of meat loaded into the tub by the operators. However, this didn't work as a quick fix, as the smaller amount of meat dumped into the grinders could not keep up with the speed of the production line. A maintenance worker replaced a broken pressure gauge and found that the hydraulic-system pressure maxed out at only 1,700 psi, even though the relief valve was set at 3,000 psi.

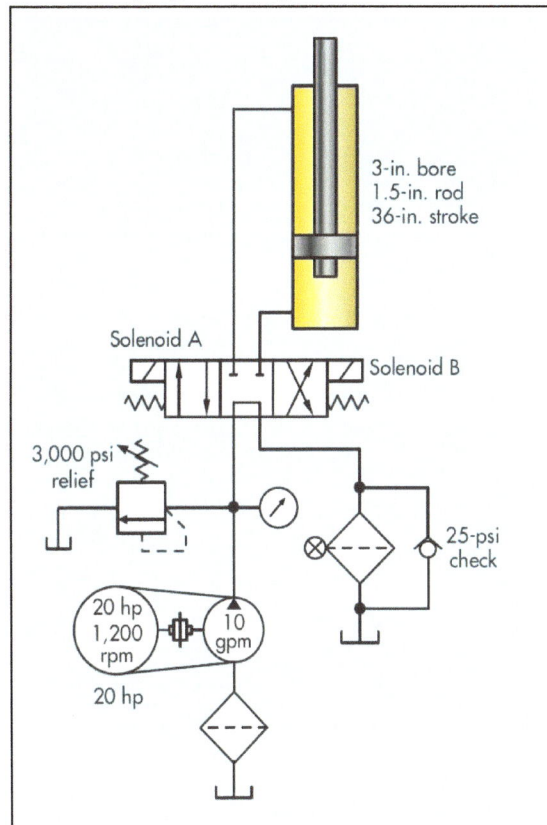

**Fig 14.27 – Hydraulic Circuit Diagram for Hot Dog Machine
(Courtesy of CFC Industrial Training)**

The 1,700 psi did not generate enough force from the cylinder to dump a full tub. However, when operators reduced the load on cylinder by only partially filling the tub, the cylinder operated at the correct speed. Of course, the first thing the maintenance people did was replace the pump, only to have the same result. They verified that the relief valve was not relieving oil to the tank at 1,700 psi and that the cylinder wasn't leaking oil from the rod-end hose. What do you think was causing the problem?

Solution:

The main cylinder in a hot dog machine could not lift tubs fully loaded with meat. Operators noticed the hydraulic system would max out at 1,700 psi, even though the relief valve was set at 3,000 psi. The pump/motor adapter had been mounted with its access hole pointed down, toward the base plate. This position blocked the view of the shaft coupling between the electric motor and hydraulic pump. Workers noticed that the motor's fan was rotating, but no one thought to check if the coupling halves were, too. They assumed the shaft key had been sheared. Installing a new key solved the problem-a part that can be had for pocket change.

14.2.18- Steel Mill

Problem: A Mill has Difficulty Moving Large Loads (Fig. 14.28):

A temper mill had a roll-removal system that operated at the proper speed but had difficulty moving large rolls off the production line. A hydraulic cylinder extended to remove the roll and was controlled by a regenerative circuit to achieve fast, slowdown, and approach speeds. The regen circuit used a pilot-operated proportional valve; its pilot connection was fed by a pressure-reducing valve that limited pilot pressure to 572 psi. The pressure at the connection had been verified. The system could move rolls weighing up to about half the rated capacity but struggled with any rolls that were heavier. They checked the cylinder piston for leakage by shutting the cap-end ball valve while the cylinder was about half-way extended. Then they pressurized the rod end of the cylinder to see if it extended, which would indicate leaking piston seals. If it locked up, the seals were good. However, you have to leak all of the supplied flow across the piston for pressure to drop-a fact overlooked by many technicians. The electrical maintenance technician was asked to examine the feedback signal' to make sure the valve spool was responding, and it was according to the technician. Any idea on why the cylinder was unable to move the heavy loads as intended?

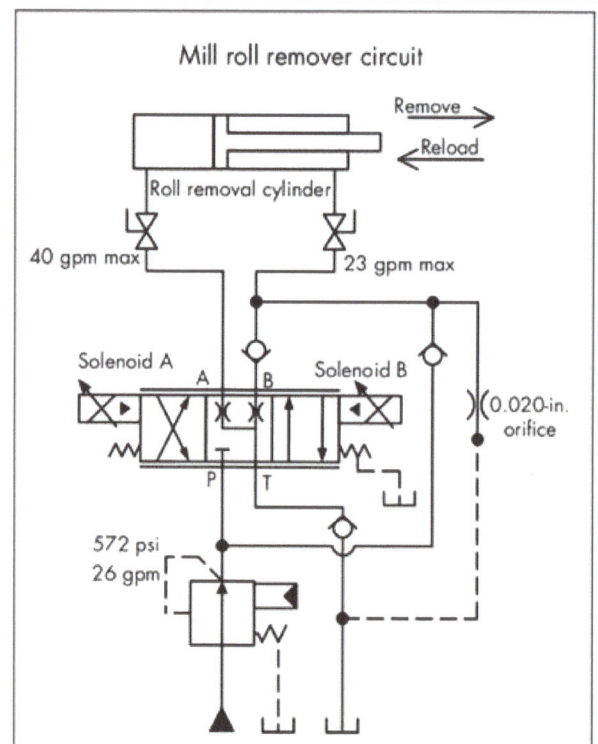

Fig 14.28 – Hydraulic Circuit Diagram for Steel Mill (Courtesy of CFC Industrial Training)

Solution:

Regenerative circuits are used to increase the speed of a cylinder by routing fluid exiting the rod end back into the cap end. However, if the circuit remains in the regenerative mode when encountering a heavy load, the cylinder will be unable to apply full force unless rod-end fluid vents to tank. The 0.020-in. orifice branching off the rod plumbing was used to accomplish this bleed-down. They found this orifice plugged and once it was cleaned, the system worked well.

14.3- Case Studies Using Analytical Fault Detection Methodology for Mobile Applications

14.3.1- Post-Hole Hammer

Problem: Post-Hole Hammer Undergoes Pump Failure (Fig. 14.29):
A fence contractor had a hydraulic *hammer* attachment for his front-end loader that hammered fence posts into the ground. He would only get about two to three months of operation from the system before the hammer's pressure-compensated pump would fail. The pump manufacturer denied warranty claims, citing "lack of lubrication" as the cause of the failure. The contractor asked the pump manufacturer what evidence led them to conclude that ran failure was due to lack of lubrication. He told them he was using the recommended oil, changed the return filters whenever the dirt indicator (ΔP) gauge approached the yellow zone, and claimed the gauge never reached the red zone. The pump was driven off a gas engine that idled at 3600 rpm and about 2200 rpm under load. They told him the piston barrel face was "welded" to the front plate it rotated against. The front cover of the pump. a Vickers PVB series also acted as the pressure plate that the barrel spins against. Any idea what was causing the premature pump failures?

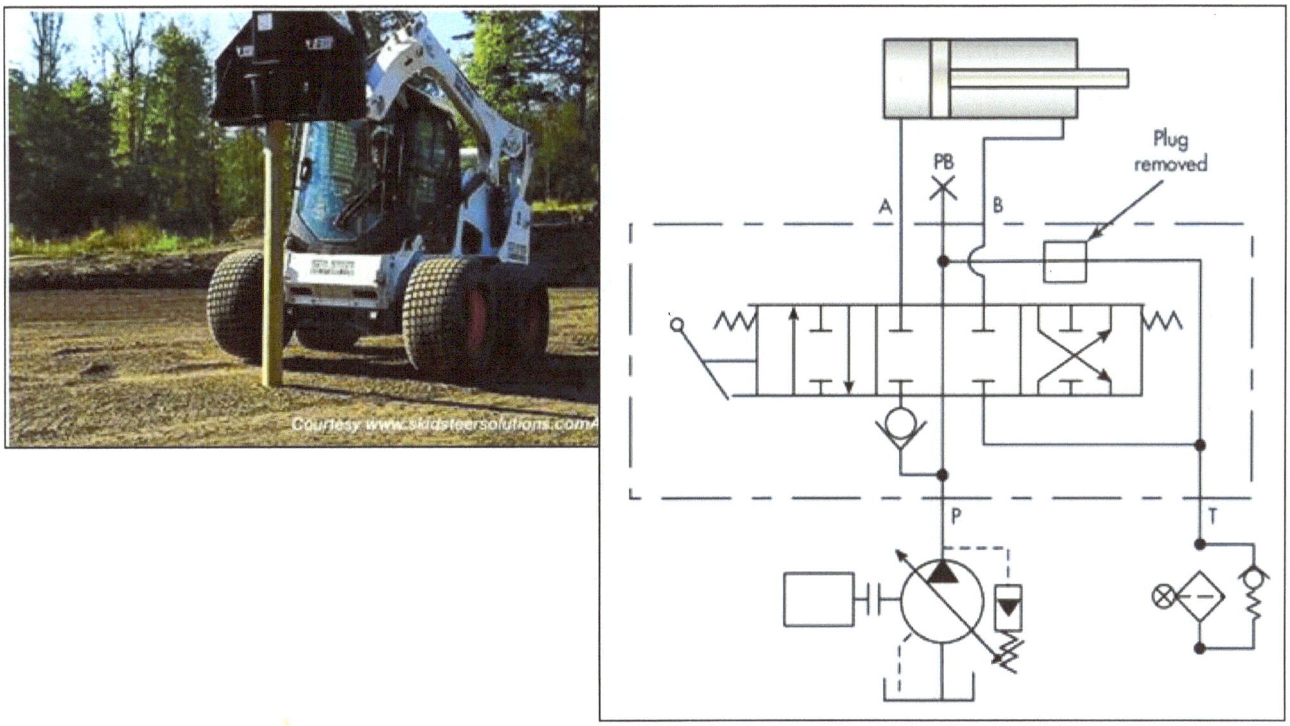

**Fig 14.29 – Hydraulic Circuit Diagram for Post Hole Hammer
(Courtesy of CFC Industrial Training)**

Solution:

The circuit was originally set up as a blocked center mobile type of directional valve with the power-beyond connection plugged. The machine owner was familiar with the power-beyond function and felt the pump did not need to compensate at 2,000 psi when idling. He had modified the power-beyond plug, essentially converting the valve to an open-center configuration. Therefore, the pump unloaded each and every time the directional valve's spool centered in neutral position. This configuration also made it easier to start the gas engine that was driving the system. Piston pumps need at least 200 to 250 psi of pressure when idling so that oil can leak into the case to lubricate the many tight-fitted moving parts. The high speed exacerbated the problem. In my opinion, this design was not a good application for a piston pump.

14.3.2- Forklift

Problem: Leaking Cylinder:

Several years ago, when I owned a large hydraulic repair shop, we received defective power steering cylinders off of a popular Clark *forklift* truck to be repaired. The Clark dealer would give us eight to 12 cylinders at a time. Some leaked externally around the rod seal area, some had large dents in the outer tube, and some were scarred internally. This required replacing the inner tube, piston, or both. We mounted the cylinders in a lathe and cut through the weld to remove the blind endcap so we could remove the rod and piston assembly. After repairing or replacing parts, we reassembled, welded, and tested the assemblies, usually in batches of eight to 12 units. I always insisted that nothing was allowed to leave our shop without testing, documentation, and stamping our job number on the repaired item. We subsequently received some complaints and warranty returns for leaking rod seals. However, when we tested the cylinders to see where they leaked, we could not get them to leak at any pressure. We would give the customer a rebuilt exchange at no charge and put the questionable cylinders in with the next batch to be repaired. The dealer called one day and insisted we go to one of his customers who purchased one of our "leaking" cylinders and installed it on his truck. When I arrived, the customer showed me a puddle of oil on the floor and wanted us to fix the problem. I cleaned up the oil mess and wiped the cylinder down and asked their fork truck operator to drive around doing his job of moving stock for about a half hour and come back to where I was waiting. I then inspected the truck and could not find any oil leaking anywhere-but there sure was a puddle on the floor when I arrived. The cylinder seemed to leak overnight, but not when the lift truck was being used. What do you think was the source of the mystery leak?

Solution:

After finding A puddle of oi under a fork 1ift-even though no leak was found after a test run-I called in our seal provider. After we explained the circumstances, he told us, "The hardest thing to seal is a cylinder with no pressure in it." That was why we could not get the returned cylinders to leak when we tested them, and why the cylinder didn't leak in the test run. Our seal guy recommended that we stop using the factory seal kit the dealer was supplying us. Instead, we should use a different seal design made from a slightly softer material. It solved our leaking rod seal problem from that point on.

14.3.3- Cargo Ship

Problem: Cargo Ship Pressure Buildup Problem (Fig. 14.30):

A *cargo ship* has a hydraulic power unit (HPU) that opens and closes several hatch doors on grain-storage areas. The accompanying schematic shows a circuit for one of eight cylinders and valves located close to the doors, with the HPU on a lower deck. The HPU is located midway between two sets of four. The pressure and return lines run approximately 200 ft. in each direction. Every time operators wanted to open or close a door; it would take almost a minute holding the solenoid button before the pressure would reach a pressure that would start moving the cylinders. Once started, they moved as expected. However, releasing the button and pushing it again caused a one-minute delay before anything would move again. The electrical system was simple. Pressing a button for any solenoid would also energize the solenoid on the unloading valve located on top of the main relief. Bleeding air from the system did slightly reduce the delay. However, can you tell the main cause of the problem from the schematic?

Solution:

Solenoid-operated relief valve typically contain three orifices. The main pressure line flows through an orifice, A in the drawing. From there, oil either leaks to the top of the main poppet, causing it to close, or drains to tank, allowing the main poppet to open at 4-bar of pressure. This orifice normally determines the closing speed of the valve. A partial blockage slows pressure buildup. The orifice acts like a flow control that controls the main poppet closing speed. Technicians found Teflon tape obstructing the orifice, thereby causing the delayed closing speed. Removing the obstruction solved the problem.

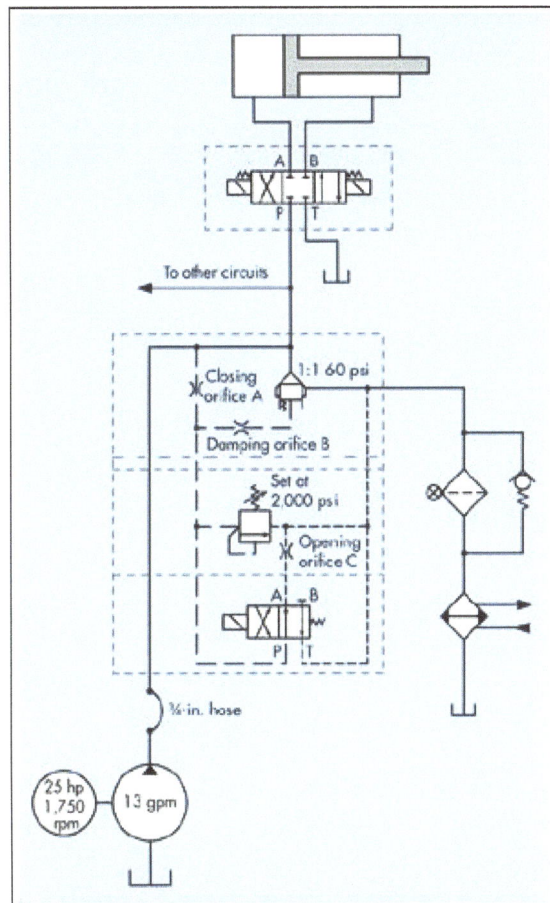

Fig 14.30 – Hydraulic Circuit Diagram for Lumberyard Stacker (Courtesy of CFC Industrial Training)

14.3.4- Lumberyard Stacker

Problem: Lumberyard Stacker Cylinder Drops (Fig. 14.31)

A *lumber* company had an older yard stacker machine that they wanted to put back into service. It had been out of service and parked due to a problem with its boom sometimes dropping. The boom lift cylinder would drop when the operator also rotated the forks at the same time to align the 4x8 and 4x12 sheets of plywood and plasterboard when placing them on top of other stacks.

I looked at the problem and noticed the operators would lift and rotate at the same time when the problem happened. If they only ran one function. the problem would not occur. All the other trucks worked well running both at the same time and expedited the unloading of the supply trucks. I told them they could use the truck but not to rotate while they were lifting. This sounds well and good, but the operators were accustomed to using the other trucks, so the operators sometimes ran lift and rotation at the same time out of habit. As a result, the problem still occurred.

The attached circuit seems to be how the hydraulic system was designed. Any idea about how to solve the problem?

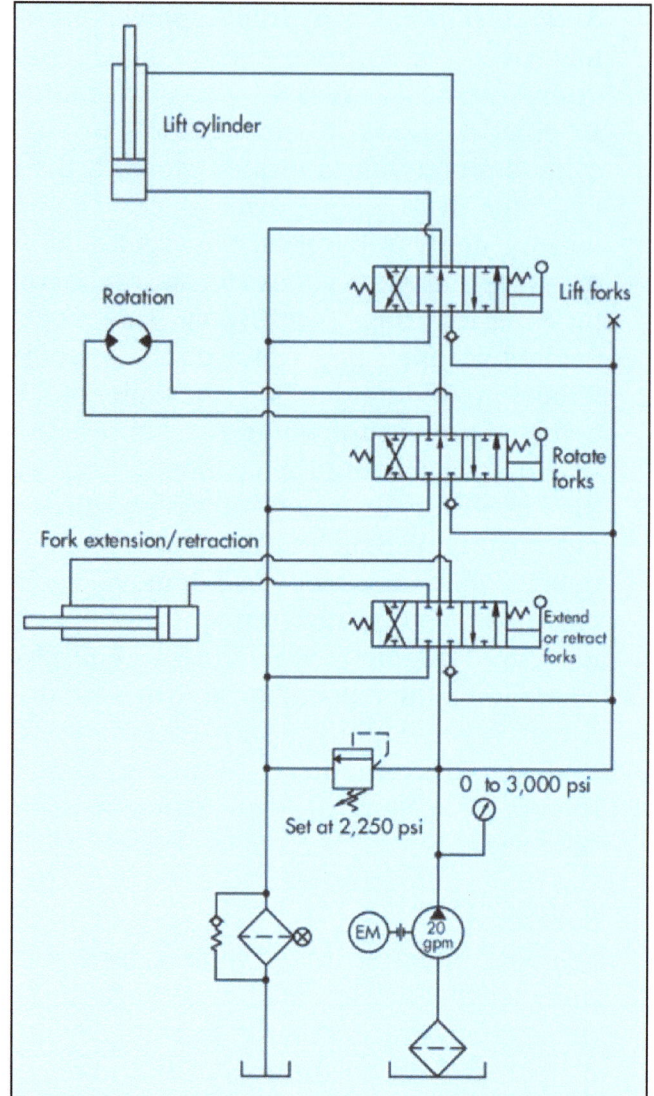

Fig 14.31 – Hydraulic Circuit Diagram for Investigated Case (Courtesy of CFC Industrial Training)

Solution:

When the yard stacker lifts suddenly dropped, it always happened when the rotation function was selected at the same time. The directional valve sections had load checks in each function. They prevented the backflow of oil from one heavier load to a lighter load function if both were shifted at the same time. We found the load check for the lift function was being held open by what looked like a roll pin. We never found out where the pin came from, but removing the obstruction fixed the problem.

14.3.5- Digger Derrick Truck

Problem: Out Riggers Locked-Up (Fig. 14.32)

A power and light company's *Digger* Derrick crew was assigned to drill several holes for power poles in an industrial complex. They arrived around 9 a.m. on a cloudy, misty, and rainy 40°F day. They set up and drilled their first two holes without having to move the truck. They then decided to break for lunch before moving on to drill the third hole since the cloud cover and misty rain gave way to a warm sun. They folded the unit up to move to the next hole location and could not get any of the outriggers to retract. All the other functions worked well, but all four outriggers seemed to be locked up. The crew called back to the maintenance department and explained the problem. The mechanic on duty told them to move the digger boom so it would extend perpendicular to a side relieving the load on the opposite side outriggers. Then he told them to see if the two outriggers opposite the boom would retract. This technique would work about 70% of the time. Operator crews should run outriggers out far enough to stabilize the rig but avoid bottoming them out. When the outriggers are bottomed out, many times they see a pressure spike caused by the pressure-compensated pump's slow response time, and the pilot-to-open load-holding check valves can have up to twice the system pressure trapped between the cylinder and PO check. The pilot-to-open ratio is too low to open the check with normal system pressure. The crew was not successful at retracting the outriggers, and when the mechanic arrived, he thanked the crew for getting him out of the garage on what turned out to be a sunny 90°F day. Without breaking a cylinder line or jacking the truck up, he performed his magic and had the truck outriggers fully retracted in less than 30 minutes. Any idea what he did to solve the problem?

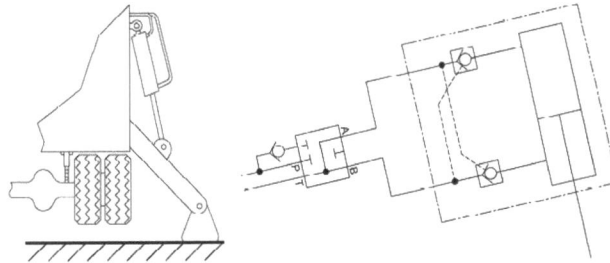

**Fig 14.32 – Hydraulic Circuit Diagram for Digger Derrick
(Courtesy of CFC Industrial Training)**

Solution

When outriggers fail to retract, not only can pressure spikes cause the problem, but temperature changes can also affect the pressure in trapped fluids. For every 10°F change in temperature, common hydraulic oil will expand approximately 0.5%. If it is trapped in a confined area like outriggers that are bottomed out, its pressure will increase approximately 1,000 psi per 10°F of temperature change. The ambient temperature (air temperature) changed 50°, increasing the trapped pressure an additional 4,000 to 5,000 psi. The mechanic brought four bags of ice from their ice-making machine, layed them on the outrigger cylinder tubes, and in no time, the oil cooled down, lowering the trapped pressure to a point where the system pressure would now open the locking pilot to open checks.

APPENDIXES

APPENDIX A: LIST OF FIGURES

Chapter 4: Troubleshooting and Failure Analysis of Pumps

Chapter 5: Troubleshooting and Failure Analysis of Motors

Chapter 6: Troubleshooting and Failure Analysis of Cylinders

Chapter 14: Examples of Hydraulic Systems Troubleshooting

APPENDIX B: LIST OF TABLES

APPENDIX C: LIST OF TROUBLESHOOTING CHARTS

T-System-01: Fluid Aeration.
T-System-02: Pump Cavitation.
T-System-03: Excessive System Noise & Vibration.
T-System-04: Excessive System Heat.
T-System-05: Low Power System.
T-System-06: Faulty System Sequence.
T-System-07: External Leakage.
T-System-08: Troubleshooting of Open Hydraulic Circuit.
T-System-09: Troubleshooting of Closed Hydraulic Circuit (Hydrostatic Transmission).
T-System-10: Actuator Slow Performance.
T-System-11: Actuator Fast Performance.
T-System-12: Actuator Erratic Performance.
T-System-13: Actuator Moves in Wrong Direction.
T-System-14: Actuator Stops to Move.
T-System-15: Actuator Load Drifts.
T-System-16: Actuator Leaks.
T-Unit-01: General Check
T-Unit-02: Noisy Unit.
T-Unit-03: Excessively Hot Unit.
T-Seal-01: Seal Troubleshooting.
T-Pump-01: No Flow out of the Pump.
T-Pump-02: Low Flow out of the Pump.
T-Pump-03: Erratic Flow out of the Pump.
T-Pump-04: Excessive Flow out of the Pump.
T-Pump-05: No Pressure at the Pump Outlet.
T-Pump-06: Low Pressure at the Pump Outlet.
T-Pump-07: Erratic Pressure at the Pump Outlet.
T-Pump-08: Excessive Pressure at the Pump Outlet.
T-Pump-09: Leaking Pump.
T-Pump-10: Excessive Pump Wear.
T-Pump-11: Air Leaks into Pump.
T-Pump-12: Excessive Pump Noise and Vibration.
T-Valve-01: DCV Troubleshooting.
T-Valve-02: FCV Troubleshooting.
T-Valve-03: PCV Troubleshooting.
T-Valve-04: EH Valve Troubleshooting.
T-Valve-05: General Valve Troubleshooting.
T-Motor-01: Motor Troubleshooting.
T-Cylinder-01: Cylinder Troubleshooting.
T-Accumulator-01: Accumulator Troubleshooting.
T-Reservoir-01: Reservoir Troubleshooting.
T-Transmission Line-01: Transmission Line Troubleshooting.
T-Heat Exchanger-01: Heat Exchanger Troubleshooting.
T-Filter-01: Filter Troubleshooting.

APPENDIX D: LIST OF REFERENCES

Hydraulic Systems Volume 1- Introduction to Hydraulics for Industry Professionals
Author: Dr. Medhat Kamel Bahr Khalil, 2016.
Publisher: Compudraulic, USA.
ISBN 978-0-692-62236-0

Hydraulic Systems Volume 2- Electro-Hydraulic Components and Systems
Author: Dr. Medhat Kamel Bahr Khalil, 2016.
Publisher: Compudraulic, USA.
ISBN: 978-0-9977634-2-3

Hydraulic Systems Volume 3- Hydraulic Fluids and Contamination Control
Author: Dr. Medhat Kamel Bahr Khalil, 2016.
Publisher: Compudraulic, USA.
ISBN: 978-0-9977816-3-2

Hydraulic Systems Volume 7- Modeling and Simulation for Application Engineers
Author: Dr. Medhat Kamel Bahr Khalil, 2016.
Publisher: Compudraulic, USA.
ISBN: 978-0-9977816-3-2

R01- Basic Electronics for Hydraulic Motion Control
Author: Jack L. Johnson, PE 1992.
Publisher: Penton Publishing Inc. 1100 Superior Avenue. Cleveland, OH 44114.
ISBN No. 0-932905-07-2.

R02- Closed Loop Electro-hydraulics Systems Manual
Author: Vickers/Eaton.
Publisher: Vickers Inc. 1992.
Training Center, 2730 Research Drive, Rochester Hills, MI 48309-3570.
ISBN 0-9634162-1-9

R03- Bosch Automation Technology
Author: Werner Gotz, Steffen Haack, Ralph Mertlick.
Publisher: Bosch.
ISBN 3-933698-05-7.

R04- Electrohydraulic Proportional and Control Systems
Publisher: Bosch Automation 1999.
ISBN 0-7680-0538-8.

R05- Proportional and Servo Valve Technology – The Hydraulic Trainer Volume 2
Author: R. Edwards, J. Hunter, D. Kretz, F. Liedhegener, W. Schenkel, A. Schmitt.
Publisher: Mannesman Rexroth AG 1988. D-8770 Lohr a. Main. ISBN   3-8023-0266-4.

R06- Proportional Hydraulics
Author: D. Scholz.
Publisher: Festo Didactic KG, Esslingen, Germany.

R07- Electricity, Fluid Power, and Mechanical Systems for Industrial Maintenance
Author: Thomas Kissell.
Publisher: Prentice Hall, Inc. 1999, Upper Saddle River, NJ 07458.
ISBN 0-13-896473-4.

R08- Fluid Power in Plant and Field – First Edition
Author: Charles S. Hedges, R.C. Womack.
Publisher: Womack Machine Supply Co. 1968.
Womack Educational Publication, 2010 Shea Road, Dallas, TX 75235.
ISBN   68-22573 (Library of Congress Card Catalog No.).

R09- Hydraulics, Fundamentals of Service
Author: Deere and Company.
Publisher: John Deere Publishing 1999.
Almon TIAC Bldg. Suite 104, 1300-19th Street, East Moline, IL 61244.
ISBN 0-86691-265-7.

R10- Industrial Hydraulics Troubleshooting
Author: James E. Anders, Sr.
Publisher: McGraw-Hill, Inc.
ISBN 0-07-001592-9.

R11- Power Hydraulics
Author: John Ashby.
Publisher: Prentice Hall 1989. Prentice Hall International, (UK) Ltd.
66 Wood Lane End, Hemel Hempstead, Hertfordshire, HP2 4RG.
ISBN 0-13-687443-6.

R12- Fluid Power with Application
Author: Anthony Esposito.
Publisher: Prentice Hall.
ISBN 0-13-060899-8.

R13- Hydraulic Component Design and Selection
Author: E.C. Fitch.
Publisher: BarDyne Inc. 5111 North Perkins Rd. Stillwater, OK 74075.
ISBN 0-9705922-3-X.

R14- Planning and Design of Hydraulic Power Systems – The Hydraulic Trainer, Vol. 3
Author: Mannesmann Rexroth GmbH.
Publisher: Mannesman Rexroth AG 1988.
D-97813 Lhr a. Main, Jahnsrtrabe 3-5 D-97816 Lohr a. Main.
ISBN   3-8023-0266-4.

R15- Logic Element Technology: Hydraulic Trainer, Volume 4
Author: Mannesmann Rexroth GmbH.
Publisher: Mannesmann Rexroth GmbH 1989.
.Postfach 340, D 8770 Lohr am Main, Telefon (09352) 180.
ISBN 3-8023-0291-5.

R16- Hydrostatic Drives with Control of the Secondary Unit. The Hydraulic Trainer, Volume 6
Author: Dr. Alfred Feuser, Rolf Kordak, Gerold Liebler.
Publisher: Mannesmann Rexroth GmbH 1989.
Postfach 340, D 8770 Lohr am Main.

R17- Control Strategies for Dynamic Systems: Design and Implementation
Author: John H. Lumkes, Jr.
Publisher: Marcel Dekker, Inc. 2002.
Marcel Dekker, Inc. 270 Madison Avenue, New York, NY 10016.
ISBN 0-8247-0661-7.

R18- Feedback Control Of Dynamic Systems
Author: Gene F. Franklin, J. David Powell, Abbas Emami-Naeini.
Publisher: Prentice-Hall, Inc.
Upper Saddle River, New Jersey.
ISBN 0-13-032393-4.

R19- Modeling and Analysis of Dynamic Systems
Author: Charles M. Close, Dean. Frederick
Rensselaer Polytechnic Institute
Publisher: John Wiley & Sons, Inc.
ISBN 0-471-12517-2.

R20- Design of Electrohydraulic Systems for Industrial Motion Control
Author: Jack L. Johnson, PE.
Milwaukee School of Engineering.
Publisher: Parker.
Copyright © Jack L. Johnson, PE 1991.

R21- Basic Pneumatics
Author: Kjell Evensen & Jul Ruud.
Publisher: AB Mecmann Stockholm 1991.
S-125 81 Stockholm, Sweden.
ISBN 91-85800*21-X.

R22- Basic Pneumatics: The Pneumatic Trainer, Volume 1
Author: Ing. –Buro J.P. Hasebrink.
D7761 Moos.
Editor: Mannesmann Rexroth Pneumatik GmbH.
Bartweg 13, W 3000 Hannover 91.

R23- Electro-Pneumatics: The Pneumatic Trainer, Volume 2
Author: Rolf Balla.
Publisher: Mannesmann Rexroth 1990, Pneumatik GmbH.
Publication No: RE 00 262/01.92.

R24- Pneumatics Theory and Applications
Author: Bosch Automation.
Publisher: Robert Bosch GmbH 1998.
Automation Technology Division, Training (AT/VSZ)
ISBN 1-85226-135-8.

R25- Fluid Power Engineering
Author: M. Galal Rabie.
Publisher: McGraw-Hill.
ISBN 978-0-07-162246-2.

R26- Air Motors Ideas with Air
Author: GAST Mfg. Co.
Publisher: GAST Mfg. Co. 1978.
P.O. Box 97, Benton Harbor, MI 49022.
Book No: Booklet #100.

R27- Air Motor Handbook
Author: GAST Mfg. Co.
Publisher: GAST Mfg. Co. 1978.
P.O. Box 117, Benton Harbor, MI 49022.

R28- Troubleshooting Hydraulic Components: Using Leakage Path Analysis Methods
Author: Rory S. McLaren.
Publisher: Rory McLaren Fluid Power Training 1993.
562 East 7200 South, Salt Lake City, UT 84171.
ISBN No. 0-9639619-1-8.

R29- Hydraulics Theory and Application From Bosch
Author: Werner Gotz.
Publisher: Robert Bosch GmbH.
Hydraulics Division K6, Postfach 30 02 40, D-7000 Stuttgart 30.
Federal Republic of Germany, Technical Publications Department, K6/VKD2.

R30- A Complete Guide to ISO and ANSI Fluid Power Symbols
Author: Fluid Power Training Institute.
Publisher: Fluid Power Training Institute 200.
562 East Fort Union Boulevard, Midvale, Utah 84047.

R31- How to Work Safely with Hydraulics
Author: Fluid Power Training Institute.
Publisher: Fluid Power Training Institute 2004.
562 East7200 South, Midvale, Utah 84047.

R32- How to Interpret Fluid Power Symbols
Author: Rory S. McLaren.
Publisher: Fluid Power Training Institute.
Rory S. McLaren 1995.
ISBN 0-9639619-2-6.

R33- Safe Hydraulics
Editor: Gates Rubber Company.
Copyright 1995.
Denver, CO 80217.

R34- Electronically Controlled Proportional Valves. Selection and Application
Author: Michael J. Tonyan.
Publisher: Marcel Dekker, Inc. 1985.
Marcel Dekker, Inc., 270 Madison Avenue, New York, NY 10016.
ISBN 0-8247-7431-0.

R35- Introduction to Closed-Loop Oil Systems
Author: Rory S. McLaren.
Publisher: Rory McLaren Fluid Power Training Institute.
7050 Cherry Tree Lane, P.O. Box 711201, Salt Lake City, UT 84171.

R36- Industrial Hydraulic Technology, Second Edition
Author: Parker Hannifin Corporation.
Publisher: Parker Hannifin Corporation 1997.
6035 Parkland Blvd, Cleveland, OH 44124-4141.
Publication No: Bulletin 0231-B1.

R37- Basic Principle and Components of Fluid Technology – The Hydraulic Trainer, Volume 1
Author: Mannesman Rexroth.
Publisher: Mannesman Rexroth AG 1988.
D-97813 Lhr a. Main, Jahnsrtrabe 3-5 D-97816 Lohr a. Main. ISBN 3-8023-0266-4.

R38- Safe-T-Bleed Corporation Catalog
Publisher: Safe-T-Bleed Corporation 2001.
Catalog No. STB-PC-1201-1

R39- Industrial Hydraulics Manual – EATON
Publisher: Eaton Fluid Power Training.
ISBN: 0-9788022-0-9.

R40- Vickers-Mobile Hydraulic Manual – Fourth Edition 1998
Author: Vickers.
Publisher: Vickers Inc. 1999.
Training Center, 2730 Research Drive, Rochester Hills, MI 48309-3570.
ISBN No. 0-9634162-5-1.

R41- Industrial Fluid Power Text, Volume 2
Author: Charles S. Hedges, R.C. Womack.
Publisher: Womack Machine Supply Company 1972.
Womack Educational Publications, 2010 Shea Road, Dallas, TX 75235.
ISBN 66-28254 (Library of Congress Card Catalog No.).

R42- Fluid Power Hydraulics and Pneumatics
Author: R. Daines.
Publisher: The Good-heart Willcox Company, Inc.

R43- Hydraulics in Industrial and Mobile Applications
Publisher: ASSOFLUID, Italian Association of Manufacturing and Trading Companies in Fluid
Power Equipment and Components

R44- Fluid Power in Plant and Field – Second Edition
Author: Charles S. Hedges, R.C. Womack.
Publisher: Womack Machine Supply Co. 1968.
Womack Educational Publication, 2010 Shea Road, Dallas, TX 75235.
ISBN 68-22573.

R45- Mobile Hydraulics Manual
Author: Eaton.
Publisher: Eaton Corporation Training.
Eden Prairie, Minnesota.
ISBN 0-9634162-5-1.

R46- EH Control Systems
Author: F.D. Norvelle.

R47- Fluid Power Journal
Publisher: International Fluid Power Society.

R48- Fundamentals of Industrial Controls and Automation
Author: Lonnie L. Smith and Mike J. Rowlett.
Publisher: Womack Educational Publications.
Dallas, Texas.
ISBN: 0-943719-04-6.

R49- Lightning Reference Handbook
Publisher: Berendsen Fluid Power.

R50- Pneumatics Basic Level
Author: P. Croser, F. Ebel.
Publisher: Festo Didactic GmbH & Co.

R51- Electro-pneumatics Basic Level
Author: F. Ebel, G. Prede, D. Scholz.
Publisher: Festo Didactic GmbH & Co.

R52- Mechanical System Components
Author: James F. Thorpe.
Publisher: Allyn and Bacon.
Needham Heights, Massachusetts.
ISBN: 0-205-11713-9.

R53- Electrical Motor Controls for Integrated Systems, Third Edition
Author: Gary J. Rockis, Glen A. Mazur.
Publisher: American Technical Publishers, Inc.
ISBN: 0-8269-1207-9.

R54- Instrumentation, Fourth Edition
Author Franklyn W. Kirk, Thomas A. Weedon, Philip Kirk.
Publisher American Technical Publishers, Inc.
ISBN: 0-8269-3423-4.

R55- Introduction to Mechatronics and Measurement Systems, Second Edition
Author David G. Alciatore, Michael B. Histand.
Publisher McGraw-Hill, Inc.
ISBN: 0-07-240241-5.

R56- Study Guides for IFPS Certification

R57- Work Books from Coastal Training Technologies

R58- Industrial Hydraulic Manual – Fourth Edition 1999
Author: Vickers.
Publisher: Vickers Inc. 1999.
Training center, 2730 Research Drive, Rochester hills, Michigan 48309-3570.
ISBN 0-9634162-0-0.

R59- Industrial Automation and Process Control
Author: John Stenerson.
Publisher: Prentice Hall.
ISBN 0-13-033030-2.

R60- Industrial Automated Systems
Author: Terry Bartelt.
Publisher: Delmar Cengage Learning.
ISBN: 10-1-4354-888-1.

R61- Introduction to Fluid Power
Author: James L. Johnson.
Publisher: Delmar Cengage Learning.
ISBN: 10-0-7668-2365-2.

R62- Summary for Engineers
Author: Dr. Abdel Nasser Zayed.
Publisher: Dr. Abdel Nasser Zayed .
ISBN: 977-03-0647-9.

R63- Mechanics of Materials
Author: Ferdinand P.Beer, E. Russell Johnston Jr., John T DeWolf.
Publisher: McGraw Hill Publishing .
ISBN: 0-07-365935-5.

R64- Oil Hydraulic System, Principles and Maintenance
Author: S. R. Majumdar.
Publisher: McGraw Hill.
ISBN 10: -0-07-140669-7.

R65- Contamination Control in Hydraulic and Lubricating Systems
Publisher: Pall

R66- Diagnosing Hydraulic Pump Failure
Publisher: Caterpillar.

R67- Oil Service Products Catalog
Publisher: Schroder Industries.

R68- Industrial Fluid Power Volume 1
Author: Charles S. Hedges.
Publisher: Womack Educational Publication.
ISBN: 0-9605644-5-4.

R69- Industrial Fluid Power Volume 2
Author: Charles S. Hedges.
Publisher: Womack Educational Publication.
ISBN: 0-943719-01-1.

R70- Industrial Fluid Power Volume 3
Author: Charles S. Hedges.
Publisher: Womack Educational Publication.
ISBN: 0-943719-00-3.

R71- Electrical Control of Fluid Power
Author: Charles S. Hedges.
Publisher: Womack Educational Publication.
ISBN 0-9605644-9-7.

R72- Hydraulic Cartridge Valve Technology
Author: John J. Pippenger, P.E.
Publisher: Amalgam Publishing Company.
Post Office Box 617, Jenks, OK 74037 USA.
ISBN: 0-929276-01-9.

R73- Noise Control of Hydraulic Machinery
Author: Stan Skaistis.
Publisher: Marcel Dekker, 270 Madison Avenue, New York, NY 10016.
ISBN: 0-8247-7934-7.

R74-Solenoid Valves
Author: Hydraforce

R75-HF Proportional Valve Manual
Author: Hydraforce

R76-Automatic Control for Mechanical Engineers
Author: M. Galal Rabie, Professor of Mechanical Engineering
ISBN: 977-17-9869-3,2010.

R77-Fluid Power System Dynamics
Author: W. Durfee, Z. Sun

Index

W

Y

www.ingramcontent.com/pod-product-compliance
Lightning Source LLC
Chambersburg PA
CBHW051600190326
41458CB00029B/6490